Springer Geochemistry/Mineralogy

More information about this series at http://www.springer.com/series/10171

Xiande Xie · Ming Chen

Suizhou Meteorite: Mineralogy and Shock Metamorphism

GUANGDONG SCIENCE & TECHNOLOGY PRESS
GUANGZHOU,P.R.C.

Xiande Xie
Guangzhou Institute of Geochemistry
Chinese Academy of Sciences
Guangzhou, Guangdong
China

Ming Chen
Guangzhou Institute of Geochemistry
Chinese Academy of Sciences
Guangzhou, Guangdong
China

NSFC Research Project; GDFPOSTL Supported Project

ISSN 2194-3176   ISSN 2194-3184   (electronic)
Springer Geochemistry/Mineralogy
ISBN 978-3-662-48477-7   ISBN 978-3-662-48479-1   (eBook)
DOI 10.1007/978-3-662-48479-1

Library of Congress Control Number: 2015950454

Jointly published with Guangdong Science & Technology Press Co., Ltd., Guangzhou, China

Springer Heidelberg New York Dordrecht London

Printed on acid-free paper

Springer-Verlag GmbH Berlin Heidelberg is part of Springer Science+Business Media
(www.springer.com)

Preface

The Suizhou meteorite is a stone meteorite, which fell on April 15, 1986, in Dayanpo, 12.5 km in the southeast of the Suizhou city, Hubei province, China. Right after the fall of this meteorite, a group of scientists from the China University of Geosciences and the Institute of Geochemistry, Chinese Academy of Sciences, conducted field survey and collection of Suizhou meteorite samples. A total weight of 270 kg of this meteorite was collected. The largest piece, a fragment of 56 kg in weight, is now preserved in the City Museum of Suizhou, and the smallest piece only weighs 20 g. This group of scientists headed by Professor Wang Renjing and Li Zhaohui then conducted a series of study on collected samples and published a monograph entitled "The synthesized study of the Suizhou meteorite" (in Chinese with an English abstract) in 1990. They classified this meteorite as an L6 chondrite on the basis of chemical composition and petrologic features, and evaluated it as a very weakly (S2) to weakly (S3) shock-metamorphosed meteorite. The authors of that book described in detail the mineral composition of this meteorite that consists of olivine, low-Ca pyroxene, plagioclase, kamacite, taenite, troilite, whitlockite, chlorapatite, chromite, and ilmenite. They observed some thin black melt veins in this meteorite, but no high-pressure phases were found.

Our recent studies revealed that the Suizhou meteorite is a unique chondrite with specific and unusual shock-related petrological and mineralogical features. At the first glance, this meteorite did contain very weakly shocked metal and troilite (S2) and the undulatory extinction is very weak for most olivine and pyroxene (S2-S3), indicating that the shock classification of S2 to S3 determined by previous investigators for Suizhou meteorite was reasonable according to the principles of Stöffler's classification. However, our detailed study revealed that most of the constituent minerals in Suizhou unmelted chondritic rock have intact structure, some of olivine and pyroxene grains display a weak mosaic texture and usually contain abundant regular fractures and 3–4 sets of planar fractures, and some of the plagioclase grains have a reduced birefringence and contain abundant planar deformation features (PDFs), and many plagioclase grains display isotropic nature, indicating these grains have transformed to maskelynite, a melted plagioclase glass.

This implies that the main three rock-forming minerals in the Suizhou unmelted chondritic rock also experienced a moderate (S4) and strong (S5) shock compression according to the Stöffler's classification for shocked chondrites. On the other hand, the thin shock melt veins in Suizhou are filled with abundant shock-produced ringwoodite, majorite and many other high-pressure mineral phases, indicating that this part of meteorite was very strongly shocked (S6). Therefore, the shock features of the Suizhou meteorite listed above match a wide range of shock stages, namely from S2 to S6, and, hence, should cover a wide range of shock-produced high pressures from 5 to >45–90 GPa and temperatures from 600 to 1750 °C. However, the results of our recent studies allowed us to come to a conclusion that the actual shock level of the Suizhou unmelted chondritic rock could be evaluated as S4-S5, and this rock experienced a shock pressure of up to 22 GPa and temperature of 1100 °C.

Locally developed thin shock veins in Suizhou were formed at pressure of 22–24 GPa that is very close to the shock pressure of the Suizhou unmelted chondritic rock but at an elevated temperature of about 1900–2200 °C. The higher temperature in melt veins than that in the unmelted chondritic rock was achieved by localized shear-friction stress caused by the collision event. Besides, the achievements in micromineralogical investigations of the Suizhou meteorite also include the following aspects:

1. Ten high-pressure mineral phases have been discovered in the shock melt veins of the Suizhou meteorite. They are as follows: ringwoodite (the high-pressure polymorph of olivine), majorite (the high-pressure polymorph of low-calcium pyroxene), akimotoite (the ilmenite-structured $(Mg,Fe)SiO_3$), devitrified perovskite (the $CaTiO_3$-structured $(Mg,Fe)SiO_3$), lingunite (the hollandite-structured polymorph of plagioclase), tuite (the γ-$Ca_3(PO_4)_2$-structured polymorph of both whitlockite and chlorapatite), xieite (the $CaTi_2O_4$-type polymorph of chromite), CF-phase (the $CaFe_2O_4$–type polymorph of chromite), garnet (majorite–pyrope in solid solution), and magnesiowüstite $((Mg, Fe)O)$. Among them, tuite, xieite, and the CF-phase are new high-pressure minerals, and lingunite is found for the first time in its single-phase form.

2. Two types of high-pressure mineral assemblages in shock veins were identified, namely (1) the coarse-grained mineral assemblage consisted of ringwoodite, majorite, akimotoite, perovskite, lingunite, tuite, xieite, and the CF-phase which formed directly through phase transformation in solid state upon shock, and (2) the fine-grained mineral assemblage consisted of majorite–pyrope$_{ss}$ (ss-solid solution), magnesiowüstite and microcrystalline ringwoodite, which crystallized from shock-induced dense melt under pressure. FeNi metal and troilite (FeS) in the veins were molten and occur as fine eutectic FeNi–FeS intergrowths in the interstices of fine-grained high-pressure minerals.

3. According to the results of high-pressure and high-temperature melting experiments on some chondrites and peridotite, the occurrence of shock-induced high-pressure polymorphs, such as ringwoodite, majorite,

lingunite, majorite–pyrope garnet, magnesiowüstite, and tuite in shock veins constrains the peak pressure of 20–22 GPa and temperature of 1800–2000 °C. However, the presence of akimotoite, devitrified perovskite, and xieite found inside or directly adjacent to the shock veins indicates that the maximum pressure and temperature developed in the Suizhou shock veins would be about 24 GPa and 2200 °C.

4. Although the shock melt veins in Suizhou are extremely thin, all of the high-pressure polymorphs, except devitrified perovskite, occur as mineral phases of good crystallinity, no matter how fine (<0.5 μm in size) their grains are, and no glassy phases were observed in any of the veins. This implies that the duration of shock-induced high-pressure and high-temperature regime in the Suizhou shock veins should be long enough (not microseconds, but a few seconds) for phase transformation of coarse-grained minerals in solid state and crystallization of fine-grained minerals from dense melt under pressure.

5. In comparison with many other shock-vein-bearing chondritic meteorites, the Suizhou shock melt veins are the thinnest and straightest, and they contain most abundant high-pressure mineral species (up to 10 polymorphs of different silicate, phosphate, and oxide minerals). It has been found that the cooling rate of the shock melt veins in meteorites is a main factor that controls the preservation of shock-produced high-pressure polymorphs. Since the hot melt veins are embedded in the cool meteorite body, and the heat diffusion coefficient of the chondritic mass is extremely low, the thinner of the shock melt vein the greater the cooling rate of the vein, and the more species of shock-produced high-pressure polymorphs could be preserved in veins.

6. A new explanation for the unique shock-related mineralogical features of the Suizhou meteorite has been proposed in that during a moderate-to-strong shock event with intensities of S4-S5, a longer duration of the shock pressure and temperature regime in the Suizhou meteorite plays an important role in the pervasive melting of plagioclase in the unmelted part of the meteorite, as well as in the formation of abundant high-pressure mineral phases in the thin shock melt veins as the temperature locally increased high enough. It has also been assumed that the phenomenon of numerous tiny chromite inclusions in molten plagioclase can be explained by disaggretation of fracture-rich chromite grains and subsequent in-situ mixing with intruding molten plagioclase.

7. Two types of zonal polymineralic grains have been found in the Suizhou chondrite. One is of $(Mg,Fe)SiO_3$ composition, and composed by three parallel zones: the inner perovskite zone, the intermediate akimotoite zone, and the outer low-Ca pyroxene zone. The other is of $FeCr_2O_4$ composition and composed also by three zones: the inner xieite zone, the intermediate CF-phase, and the outer chromite zone. The inner zones for both grains are just adjacent to the walls of shock melt veins. Detailed studies demonstrate that the existence of temperature gradient from the vein wall to the unmelted chondritic rock, and fast cooling of the extremely thin shock veins are regarded as essential conditions for the formation of such zonal polymineralic grains in the Suizhou meteorite.

8. Some large fragments consisted of three or two high-pressure polymorphs of silicate minerals, namely ringwoodite, majorite, and lingunite, were observed in the Suizhou shock veins. Besides, the distinct two-phase grains consisted of xieite and a high-pressure polymorph of one of the above-mentioned three silicate high-pressure minerals are also observed in veins. The interfaces between high-pressure polymorphs of silicate or oxide minerals in these fragments are quite sharp, implying that partial melting does not take place at the interface areas.

9. The different behavior of FeNi metal in the Suizhou meteorite upon shock was very well explained by using the theory of local concentration of stress at the boundaries of two phases with different densities, or at the discontinuities in rocks or minerals. While the FeNi metal in the Suizhou meteorite is "very weakly" shocked and displays no obvious shock-induced intragranular textures, some of very small rounded FeNi metal grains with higher Ni content were observed in cracks or at intersecting joints of shock-induced planar fractures in olivine and pyroxene, indicating that these tiny grains must be deposited from vapor phase produced by local stress concentration during shock event.

10. The discovery of various high-pressure phases in the Suizhou shock veins is of important significance in the study of Earth's mantle mineralogy. While the minerals such as ringwoodite, majorite, lingunite, and majorite–pyrope garnet are main stable mineral phases for the Earth's mantle transition zone, the perovskite, akimotoite, magnesiowüstite, and xieite are the stable mineral phases in the P–T conditions of the lower mantle.

Mineralogy is an important basic discipline of geological sciences. Mineralogical research does not have to be ended in itself. Mineralogical research should be aimed at solving geological problems. This book introduces the unique characteristics of different minerals found both in unmelted chondritic rock and in shock-induced melt veins of the Suizhou meteorite and describes the specific shock-related mineralogical features of this meteorite in an attempt to draw some conclusions on its P–T history and to enrich the contents of mineralogy and geochemistry of the Earth's mantle. We hope that mineralogists both in foreign countries and in our homeland who are engaging in study of shock-induced mineralogical features of terrestrial or extraterrestrial rocks may find this book useful and helpful for their own undertakings.

We are indebted to the National Natural Science Foundation of China for supporting the study of Suizhou meteorite under grants 41172046, 40772030, 40272028, 49825106, and 49672098. We thank the Guangdong Foundation for Publication of Outstanding Science and Technology Literatures for its support and provision of funds for printing this book. We are particularly grateful to the Guangzhou Institute of Geochemistry, Chinese Academy of Sciences (GIGCAS), for its profound concern of studying this meteorite. We Would like to thank Professors Zhaohui Li, Xianxian Ye, Jingzhong Chen, Shangyue Shen, and to the City Museum of Suizhou for providing us the samples or thin sections of the Suizhou meteorite. Special thanks are extended to Doctors Hokwang Mao,

Yingwei Fei, Jinfu Shu, and M.E. Minitti of the Geophysical Laboratory, Carnegie Institution of Washington, for their great help in studying of the new high-pressure minerals tuite, xieite, and the $CaFe_2O_4$–structured $FeCr_2O_4$. Many thanks to Deqiang Wang, Ying Wang, Dayong Tan, and Linli Chen of GIGCAS, and Xiangping Gu of the Central South university for their help in scanning electron microscopy, X-ray and Raman microprobe, electron microprobe, and X-ray microdiffraction analyses, respectively. The first author of this book would like to express his appreciation to his wife, Professor Jie Li of the Guangdong University of Technology, for her fervently support and profound concern during the period from the beginning to the end of writing this book.

Guangzhou   Xiande Xie
March 2014   Ming Chen

Contents

Chapter 1
General Introduction of the Suizhou Meteorite

The Suizhou meteorite is a fall and was classified as an L6 chondrite. Right after the fall of this meteorite, a group of scientists from China University of Geosciences (CUG) and the Institute of Geochemistry, Chinese Academy of Sciences (IGCAS) conducted field survey and collected Suizhou meteorite samples. This group headed by Professor Wang Renjing of CUG and Professor Li Zhaohui of IGCAS then conducted a series of studies on collected samples and published a monograph entitled "A synthetical study of Suizhou meteorite" in Chinese in 1990 (Wang and Li 1990). In this chapter, we describe the results of on-the-spot investigations together with a brief introduction of the chemical composition, mineralogy, structure characteristics, isotope composition, and chronology, as well as the formation and evolution history of this meteorite. Main material of this chapter was quoted from that monograph. The authors of this book only added a few pictures of Suizhou meteorite they took in the Suizhou City Museum and some newly obtained data on high-pressure mineralogy in shock melt veins and P-T history of shock metamorphism of this meteorite.

1.1 Falling Phenomenon of the Suizhou Meteorite

On April 15, 1986, at 18:52 Beijing time, the Suizhou meteor entered the atmosphere at a relatively steep angle over Dayanpo near the Suizhou City, Hubei Province, China (Fig. 1.1), and violently exploded at the altitude of about 10,000 m, followed by the secondary explosion—fragmentation at lower altitude (Shu and Yuan 1990). During its atmospheric passage, the Suizhou meteor suffered ablation under high-temperature and low dynamic pressure conditions. As a result, ~ 31 % of the surface of the meteor was fused, forming the fusion crust 0.5–1.5 mm in thickness. At 6 p.m., this meteor formed a stone meteorite shower and then fell on the earth surface. The falling area of the Suizhou meteorite is about 10 km from east to west and 5 km from south to north, and the strewn field of elliptical form is estimated to be more than 38 km^2. Its geographical coordinates are as follows: E113° 25′–E113° 32′ and N31° 36′–N31° 38′ (Wang and Li 1990).

© Springer-Verlag Berlin Heidelberg and Guangdong Science
& Technology Press Co., Ltd. 2016
X. Xie and M. Chen, *Suizhou Meteorite: Mineralogy and Shock Metamorphism*,
Springer Geochemistry/Mineralogy, DOI 10.1007/978-3-662-48479-1_1

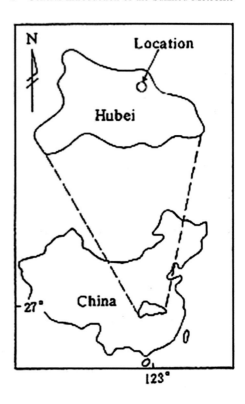

Fragments of the Suizhou meteorite shower were distributed over the strewn field with the biggest ones in the west and the smallest ones in the east part of the field. Fragments were found and collected from 72 different spots in the area. The total weight of fragments is more than 270 kg, the biggest one weighs 56 kg (Fig. 1.2), and the smallest one is only 20 g. Figure 1.3 illustrates some fragments of the Suizhou meteorite covered with black fusion crusts. The eyewitnesses of Suizhou meteorite shower introduced the following falling phenomenon (Wang and Li 1990):

1. The biggest fragment of 56 kg in weight hit the wheat field forming a round hole of 75 cm in diameter and 50 cm in depth. The meteorite fragment has round shape and covered with dark brown to black fusion crust of 0.5–1 mm in thickness.
2. A fragment of 2.35 kg in weight broke the roof tiles and a wooden rafter of the farmer Zhang's house, and hit the room cement floor, forming an elliptical pit of 19 cm × 13 cm × 5 cm in size.
3. A fragment of 5 kg in weight hit and penetrated the concrete-slab roof of the Dayanpo grain supply station and formed a hole of 25.5 cm × 20 cm in size.

Fig. 1.2 The biggest fragment of the Suizhou meteorite with weight of 56 kg

Fig. 1.3 Fragments of the Suizhou meteorite with black fusion crusts

4. A fragment of 20 g in weight broke a big roof tile of the farmer Ye's house into 4 pieces, and the meteorite fragment was crushed into small pieces of soybean or rice grain sizes.

1.2   Chemical Composition

The fresh chondritic rock of the Suizhou meteorite shows light gray color (Figs. 1.2 and 1.3). Investigators of both CUG and IGCAS selected fresh samples from the four Suizhou fragments and conducted chemical analysis for these samples separately (Huang and Xiao 1990; Hou et al. 1990). Their results are summarized in Table 1.1. From this table, we can see that the chemical compositions of four fragments taken from different locations are almost identical that implies the uniformity in composition for the Suizhou meteorite.

The main chemical parameters of the Suizhou meteorite were calculated and listed in Table 1.2 (Hou et al. 1990). These parameters are in good consistency with those for L-group chondrites; hence, the Suizhou meteorite is classified as an

Table 1.1 Bulk chemical composition of the Suizhou meteorite, wt%

Location	Dayanpo[a]	Wenjiayan[a]	Tiaoshuihe[b]	Yongfengyuan[b]	Average
SiO_2	39.64	39.77	39.20	39.80	39.60
TiO_2	0.11	0.11	0.12	0.12	0.12
Al_2O_3	2.23	2.21	2.24	2.16	2.21
FeO	15.12	14.54	15.58	15.55	15.20
MgO	24.84	24.85	24.86	24.99	24.89
CaO	1.76	1.80	1.62	1.58	1.69
MnO	0.38	0.38	0.35	0.35	0.37
Na_2O	0.93	0.94	1.08	1.06	1.00
K_2O	0.11	0.11	0.13	0.13	0.12
H_2O^+	0.00	0.00	0.038	0.039	0.019
H_2O^-	0.10	0.06	0.056	0.052	0.067
P_2O_5	0.23	0.23	0.23	0.24	0.23
Cr_2O_3	0.60	0.59	0.51	0.52	0.56
Fe^o	6.28	6.74	6.12	5.90	6.26
Ni	1.25	1.23	1.56	1.50	1.39
Co	0.065	0.064	0.058	0.056	0.061
Zn	0.035	0.012	–	–	0.023
Cu	0.043	0.013	0.009	0.008	0.018
FeS	6.42	6.61	6.47	5.98	6.37
C	0.059	0.038	–	–	0.048
Total	100.20	100.30	100.23	100.04	100.19
TFe	22.11	22.24	22.34	21.79	22.12

[a]Data from Huang and Xiao (1990); [b]Data from Hou et al. (1990)

Table 1.2 Main chemical parameters of the Suizhou meteorite

	Fe/SiO$_2$	Fe^o/Fe	Fa(%)[a]	SiO$_2$/MgO	Fe/Si[b]	Si/Mg[b]
Suizhou	0.56	0.27	23–24	1.59	0.60	1.06
L-group	0.55 ± 0.05	0.33 ± 0.07	22–26	1.59 ± 0.05	0.59 ± 0.03	1.07 ± 0.03

[a]FeO/(FeO + MgO)mol% of olivine; [b]Atomic ratio, others are weight ratio

L-group chondrite. The abundance of trace elements in Suizhou meteorite was determined by instrumental nuclear activation analysis, and the results show that the contents of 25 trace elements, such as Sc, V, Ga, As, Se, Os, Ir, Au, Sb, Br, Ru, U, Th, La, Ce, Nd, Sm, Eu, Yb, and Lu, are consistent with those for mean L6 and mean L-group chondrites (Wang 1990; Zhong et al. 1990). The atomic ratio U/Au for Suizhou = 2.71, and Br/Se = 0.006, La/Sm = 1.68, Lu/Sm = 0.16, and Th/U = 3.6 (Zhong et al. 1990). All they are consistent with those for L-group chondrites.

1.3 Mineral Composition

The Suizhou meteorite is an L6 chondrite (Wang et al. 1990a; Zeng 1990). The fresh surface of the meteorite has light gray color. There are a few black shock-induced melt veins of 0.02–0.1 mm in width which were observed in this meteorite (Fig. 1.4). These veins contain abundant dense mineral polymorphs formed at extremely high

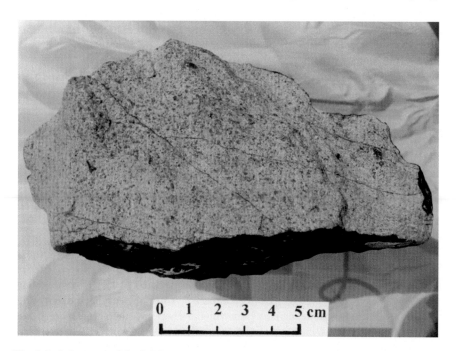

Fig. 1.4 A fragment of the Suizhou meteorite with thin shock melt veins

Fig. 1.5 A broken fragment of the Suizhou meteorite covered by black fusion crust

dynamic pressures and temperatures (Xie et al. 2001a). During its atmospheric passage, the Suizhou meteor suffered ablation under high-temperature and low dynamic pressure condition. As a result, ∼31 % of the surface of the meteor was molten, forming the fusion crust of 0.5–1.5 mm in thickness (Fig. 1.5). In this paragraph, we shall introduce the mineral compositions of Suizhou chondritic rock, as well as minerals in shock melt veins and in fusion crust.

1.3.1 Minerals in Chondritic Rock

Olivine is a main constituent mineral of the Suizhou meteorite. Most of olivine grains are in granular form. Their grain size ranges from 0.015 mm^2 × 0.01 mm^2 to 1.0 mm^2 × 0.8 mm^2. According to the results of electron microprobe analysis (EPMA) for 20 olivine grains, its Fa value is ranging from 23.9 to 25.7. The measured optical constants are as follows: Ng = 1.7195, Nm = 1.7016, Np = 1.6824, $Ng - Np$ = 0.0371, and (—)2 V = 87° (Zeng 1990).

Pyroxene is a second abundant rock-forming mineral in the Suizhou meteorite. The pyroxene grains usually have semi-idiomorphic form, and the grain size ranges from 0.02 mm^2 × 0.01 mm^2 to 0.8 mm^2 × 0.3 mm^2. EPMA of 17 pyroxene grains shows that the Fs value is ranging from 20.8 to 22.8. This indicates that the pyroxene in Suizhou is orthopyroxene, namely hypersthene. Their measured optical constants are as follows: Ng = 1.695, Nm = 1.689, Np = 1.681, $Ng - Np$ = 0.014, and (—) 2 V = 82°. EPMA also revealed two clinopyroxenes inside Suizhou chondrules: One is pigeonite with a composition of $Wo_{6.5}En_{74.5}Fs_{19.0}$ and another is diopside with a composition of $Wo_{45.3-46.5}En_{45.7-46.7}Fs_{6.8-9.0}$ (Zeng 1990).

Plagioclase occurs as grains of irregular form in the interstices of other rock-forming minerals with grain size of 0.005–0.3 mm. EPMA of 12 plagioclase grains shows that it has a composition of $Or_{6.6-12.0}Ab_{67.9-81.3}An_{11.6-20.1}$. Most of the plagioclase grains have smooth surface with first-grade gray-white interference color. Only a very few grains show polysynthetic twins (Zeng 1990). Our recent observations and micro-Raman studies revealed that most of the plagioclase grains were melted and transformed into maskelynite (Xie et al. 2001a).

FeNi metal (kamacite and taenite) in the Suizhou meteorite occurs mostly in the form of granular or irregular grains. The grain size is ranging from several microns to 80–320 μm. Chemical analysis of 19 metal grains gives an average composition of Fe = 93.72, Ni = 5.73, Co = 0.69, Cu = 0.13, Ga = 0.07, Ge = 0.50, and S = 0.04 wt%. The measured hardness of Suizhou kamacite is 185.7 kg/mm^2, and its cell parameter is a = 2.8670 Å. Taenite in Suizhou is rare and has much higher Ni content (8.21–56.06 wt%) in comparison with that of kamacite. Its average composition is Fe = 66.99, Ni = 31.84, Co = 0.32, Cu = 0.30, Ga = 0.12, Ge = 0.33, and S = 0.02 wt%. The measured hardness of taenite is 263.8 kg/mm^2, and its cell parameter is a = 3.5846 Å (Shen and Zhuang 1990).

Troilite occurs as grains of irregular form in the interstices of other rock-forming minerals with grain size ranging from several tens to several hundreds microns. EPMA of 16 troilite grains gives an average composition as Fe = 63.18, Ni = 0.30, Cu = 0.23(9), Ga = 0.14, Ge = 0.16, and S-36.03 wt%. The measured hardness of troilite is 239.3 kg/mm^2, and its cell parameters are a = 5.9546 Å and c = 11.7316 Å (Shen and Zhuang 1990).

Chromite in Suizhou meteorite shows gray color under reflected light and occurs as single semi-idiomorphic to idiomorphic crystals. Their grain size ranges from several microns to 150 microns. EPMA results of three chromite grains give average chemical composition as FeO = 30.713, MgO = 2.295, CaO = 0.072, ZnO = 0.515, MnO = 0.652, TiO_2 = 2.569, SiO_2 = 0.156, Al_2O_3 = 5.639, Cr_2O_3 = 57.078, and V_2O = 0.998 wt% (Shen and Zhuang 1990). Its chemical formula is $(Fe_{0.90}Mg_{0.12})_{1.02}(Cr_{1.59}Al_{0.24}Ti_{0.07}V_{0.02}Mn_{0.02}Zn_{0.02})_{1.95}O_4$, showing that such a composition is characteristic for chromites in meteorites and different from chromites in terrestrial rocks. The measured hardness is 699.9 kg/mm^2 (Shen and Zhuang 1990).

Ilmenite in Suizhou meteorite has a similar occurrence as chromite, but with much smaller grain size. Under reflected light, ilmenite shows weak nonisotropic feature and gray color. EPMA of 3 ilmenite grains gives an average composition as FeO = 40.125, MgO = 1.894, CaO = 0.058, ZnO = 0.026, MnO = 1.894, TiO_2 = 53.097, SiO_2 = 0.076, Al_2O_3 = 0.069, Cr_2O_3 = 0.690, and V_2O = 0.284 wt %. Its calculated chemical formula is $(Fe_{0.82}Mg_{0.14}\ Mn_{0.04})_{1.00}(Ti_{0.98}Cr_{0.01})_{0.99}O_3$ (Shen and Zhuang 1990).

Natural cupper occurs as small inclusions of irregular shape in FeNi metal, or in rare cases, in the interstices of metal, troilite, and silicates. It is isotropic and shows purplish red color in reflected light. EPMA gives the chemical composition as Fe = 5.14, Ni = 2.29, Cu = 91.63, Ga = 0.02, Ge = 0.37, and S = 0.18 wt%. Its formula is $Cu_{0.917}Fe_{0.059}Ni_{0.025}$ (Shen and Zhuang 1990).

Whitlockite in Suizhou meteorite occurs as individual grains of irregular shape in the interstices of matrix minerals. Its grain size is ranging from 0.07 to 0.3 mm. Under transmitted light, it is colorless and transparent and shows no cleavages but with some irregular fractures. Its interference color is first-grade gray-white and often shows wavy extinction. Optical parameters are as follows: uniaxial negative, No = 1.627, and Ne = 1.624. EPMA of 3 whitlockite grains in the Suizhou meteorite gives an average composition (in wt%) as FeO = 0.06, MgO = 3.27, CaO = 46.62, NiO = 0.08, Na_2O = 2.57, K_2O = 0.03, TiO_2 = 0.06, Cr_2O_3 = 0.03, and P_2O_5 = 47.67 (Xie et al. 2001a, b, c). Its formula is $(Ca_{2.52}Mg_{025}Fe_{0.01}Na_{0.25})_{3.03}(P_{1.02}O_4)_2$.

There are some other minerals which were described in several papers of the book "A Synthetical study of Suizhou meteorite," such as rutile, pyrrhotite, zircon, corundum, hornblende, gold, pyrite, chalcopyrite, arsenopyrite, galena, sphalerite, alabandite, graphite, and two species of Nb–Ta minerals with the structure of ixiolite (Shen and Ren 1990; Chen 1990; Shen 1990). All were found in the products of mechanical mineral separation of the Suizhou hand samples, and no enough identification data for these minerals were given. Therefore, we are not going to introduce them in detail. It is worth to point out that the Raman spectroscopic study on the Suizhou powder sample revealed a weak Raman peak at $1607~cm^{-1}$, which is characteristic for the line G of graphite, but line D at $1337~cm^{-1}$, the disorder scattering line of graphite, is absent (Li et al. 1990). This indicates that the degree of thermal metamorphism is so high that all amorphous carbon in the Suizhou meteorite was converted to graphite through ordering and graphitization. As estimated from the order parameter of carbonaceous material, the temperature of thermal metamorphism is $\leq777~°C$ in the position where the Suizhou meteorite was located in its parent body (Li et al. 1990).

1.3.2   Minerals in Shock Melt Veins

In 1990, a group of mineralogists studied the mineral composition of shock-produced melt veins using optical microscopy and electron microprobe techniques and observed that the melt veins are composed of glassy materials, olivine, pyroxene, plagioclase, and eutectic intergrowths of FeNi metal (taenite) and troilite (Qiu et al. 1990; Shen and Zhuang 1990). They indicated that the Suizhou melt veins experienced the shock-induced temperature as high as 900 °C (Shen and Zhuang 1990). However, they could not find any of shock-induced high-pressure minerals in the Suizhou melt veins.

Some years later, the authors of this book studied for the first time the mineral composition of two black shock melt veins of 0.02–0.09 mm in thickness in a polished thin section of the Suizhou meteorite using scanning electron microscopy, electron microprobe, and Raman spectroscopy and surprisingly found that both of

Fig. 1.6 BSE image of an enlarged view of a straight shock melt vein in the Suizhou meteorite (Xie et al. 2011a, b)

the melt veins in Suizhou are full of high-pressure minerals (Fig. 1.6). It was also found that the thin melt veins are mainly composed of a fine-grained matrix that makes up 80–90 % by volume of the veins, consisting of high-pressure minerals majorite–pyrope$_{ss}$ (ss = solid solution), magnesiowüstite, and ringwoodite, and FeNi + FeS in eutectic intergrowths. The remaining 10–20 % by volume of the veins consist of coarse-grained high-pressure phases, including ringwoodite (the γ phase of olivine), majorite (the garnet-structured pyroxene), lingunite (the hollandite-structured plagioclase), tuite (γ-Ca$_3$(PO$_4$)$_2$, the high-pressure polymorph of whitlockite), xieite (the CaTi$_2$O$_4$-structured polymorph of chromite), and the CaFe$_2$O$_4$-structured polymorph of chromite (Xie et al. 2001a, b, 2002, 2003; Chen et al. 2003a, b). The detailed description of both coarse- and fine-grained high-pressure minerals in shock melt veins of the Suizhou meteorite will be given later in Chap. 5.

1.3.3 Minerals in Fusion Crust

The fusion crust is composed mainly of glassy material with some residual or newly formed silicate minerals. FeNi metal and sulfide are distributed at the lower part of the crust, while magnetite, the oxidation product of metallic minerals, is well developed on the upper part of the crust. According to the study of Qiu et al. (1990) and Shen and Zhuang (1990), the mineralogical constituents of the Suizhou fusion crust are as follows.

Silicate glass is the main constituent of the fusion crust. It shows some cryptocrystalline features. Under the polarizing microscope, the fusion crust is semitransparent to opaque and has brown to black color. It is revealed by EPMA that 87 % (by volume) of the silicate glass showed the compositions similar to those of olivine, pyroxene, and plagioclase, respectively. Only 13 % (by volume) of the glass has the composition of mixed silicate melt. This indicates that the silicate material in fusion crust must experience rapid melting and fast quenching. The molten minerals with high viscosity solidified so quickly that they were not able to mix up totally in the crust. The small amount of mixed melt has following compositions (in wt%): SiO_2 = 45.24 and 48.27, Al_2O_3 = 2.44 and 7.39, FeO = 11.82 and 14.38, MnO = 0.41 and 0.09, MgO = 34.34 and 25.32, CaO = 1.91 and 1.31, Na_2O = 1.03 and 1.00, K_2O = 0.01 and 0.25, P_2O_5 = 0.53 and 0.25, Cr_2O_3 = 0.20 and 0.26, and NiO = 1.14 and 0.97.

Olivine in fusion crust has the chemical composition (Fa = 25–28) similar to that of olivine in Suizhou chondritic mass (Fa = 24–26). However, the average Fa for fusion crust is 27 that is a little higher than that of olivine in Suizhou chondrite ($Fa_{Avg.}$ = 24). It implies that most of olivine crystals in the fusion crust are residual grains, but there exist a small portion of olivine grains which were recrystallized from iron-rich melt.

Pyroxene in the Suizhou fusion crust has the composition of $En_{76.9}Fs_{21.5}Wo_{1.6}$, which is similar to that of pyroxene in the chondritic portion. The average Fs for pyroxene in the fusion crust is 21.8 that is a little higher than that for pyroxene in Suizhou chondrite ($Fs_{Avg.}$ = 21.3). This shows that besides the residual pyroxene grains, there also exist a few pyroxene grains that were crystallized from iron-rich melt.

Plagioclase is also observed in Suizhou fusion crust. EPMA results showed the following data for its chemical composition: Or = 18.5, Ab = 35.2, An = 46.3, and No = 56.8. Obviously, such a composition is different from that of plagioclase in Suizhou chondrite, which has the composition of Or = 6.6–15.2, Ab = 39.0–82.4, An = 11.0–45.8, and No = 11.8–54.0. Therefore, plagioclase in fusion crust is much richer in K and Ca in comparison with that in chondrite, indicating that it has two origins: Oe is remnant plagioclase, and the other is product of crystallization from K- and Ca-rich plagioclase melt.

FeNi metal and **sulfide** is also observed in Suizhou fusion crust. They occur mainly in the form of eutectic intergrowths. The fine-grained taenite is embedded in the groundmass of troilite. The chemical composition of taenite is (in wt%) Fe = 90.09–91.11, Ni = 7.93–7.99, Co = 0.44–0.54, Cu = 0.00–0.02, Ga = 0.06–0.44, Ge = 0.29–0.33, and S = 1.00–1.18. Troilite has chemical composition similar to that in chondrite. EPMA of 12 troilite grains gave its average composition (in wt%): Fe = 63.316, S = 36.018, Ni = 0.277, Co = 0.109, Cu = 0.322, Ga = 0.184, Cr = 0.086, and P = 0.022. Optical observations show that the molten metal and sulfide are concentrated at the lower part of the crust in the form of vein networks.

Magnetite occurs in the fusion crust as dendritic grains or needle-like crystals in the upper part or at the surface of the crust. It is a product of melting and oxidation of FeNi metal and other Fe-bearing minerals during the atmospheric passage of the Suizhou meteorite.

1.4 Texture Characteristics

The chondritic texture of the Suizhou meteorite can still be recognized on hand specimens and in thin sections. According to the statistics of measurements on 5 thin sections, the Suizhou meteorite contains (by volume) 75.11 % of transparent minerals, 11.66 % of opaque minerals, and 13.23 % of different types of chondrules (Wang et al. 1990b). Besides chondrules, the Suizhou meteorite shows crystalline granular texture. Some colorless transparent silicate glassy material has been recrystallized into fibrous microcrystalline aggregates. All these features imply that the Suizhou meteorite experienced high-grade thermal metamorphism, and it can be assigned as petrological type 6 chondrite (Hou et al. 1990).

1.4.1 General Characteristics of Chondrules

Small amount of chondrules can be observed in Suizhou meteorite under microscope. Most of these chondrules have spherical or elliptical forms. The grain size of chondrules varies in a wide range. The largest one has the diameter of 7.1 mm, and the smallest one is only 0.3 mm in diameter. The average diameter of Suizhou chondrules is ~ 1 mm. The main characteristics of Suizhou chondrules are as follows: (1) The glassy material in chondrules has been devitrified and recrystallized into colorless transparent matter of irregular form and distributed preferentially at the peripheries of chondrules; and (2) only a small amount of chondrules show clear boundaries with meteorite groundmass, and the outlines of most chondrules are indistinct indicating the presence of high-grade thermal metamorphism of the meteorite (Wang et al. 1990a, b).

1.4.2 Textural Types of Chondrule

On the basis of texture characteristics observed by Wang et al. (1990b), the Suizhou chondrules can be classified into following 10 basic textural types:

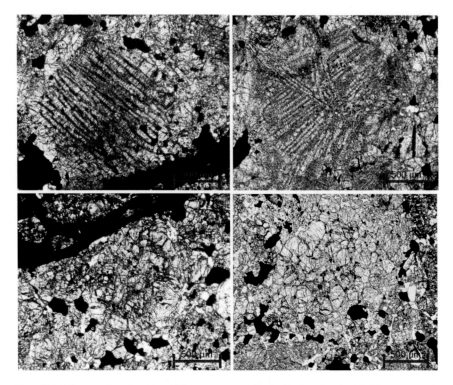

Fig. 1.7 Micrographs showing different types of chondrules in the Suizhou meteorite (Photographs taken by authors of this book)

1. Grated chondrules: The olivine single crystals arranged in a certain direction, and a ring belt of olivine formed on the margins of chondrules (Fig. 1.7 upper left). The devitrified material filled the space in between the grates of the chondrules.
2. Parallel fibrous chondrules: This type chondrule consists of only one mineral— olivine or orthopyroxene—or, in some cases, of both. Prismatic or fibrous crystals distributed in one parallel direction.
3. Radiating chondrules: The radiating center located in the center of the chondrule and the fibrous orthopyroxene or needle-like opaque minerals distributed from that center.
4. Fan-shaped chondrules: This type of chondrules has one or several radiating points responsible for mono- or multifan patterns (Fig. 1.7 upper right). Some devitrified fibrous material can be observed interstitially in pyroxene fibers.
5. Swirling and radiating chondrules: Prismatic olivine arranged in eight parallels around the margins of chondrules in the swirling and radiating arrangement within chondrules.

6. Check-like chondrules: The chondrules consist of two single olivine or pyroxene crystals with parallel striped structure, which are intersecting each other and form a check pattern.

7. Porphyritic chondrules: This type of chondrules consists of euhedral or semi-euhedral olivine porphyritic crystals with glassy material interstitially to them.

8. Micro-granular chondrules: The chondrules of this type are fully crystalline and show distinct boundaries. They mainly consist of irregular or semi-euhedral olivine or pyroxene crystals (Fig. 1.7 bottom left and bottom right). The grain size is round 0.02 mm.

9. Ring-like chondrules: The chondrules of this type are rare. They have opaque minerals in the central area and a ring consisted of colorless devitrified material.

10. Compound chondrules: These chondrules are very rare. They often contain other smaller chondrules inside themselves.

Among above-mentioned types, the micro-granular, porphyritic, parallel fibrous, and grate-like chondrules are most common. The others are rarely observed in the Suizhou meteorite.

1.5   Isotope Compositions and Chronology

The isotope compositions of lead, oxygen, and sulfur and chronology, as well as the noble gases, and fission track dating of the Suizhou meteorite have been investigated by several previous investigators (Wang et al. 1990b; Lu and Wang 1990; Wang and Li 1990). In this paragraph, we shall briefly introduce the main results of their investigations.

1.5.1   Lead, Oxygen, and Sulfur Isotope Compositions

For lead isotope analysis, the fresh specimens of Suizhou meteorite were crashed and ground into powders of different sizes. Specimens S1 (Dayanpo) and S2 (Minjiachong) were divided into two probe samples (200 mesh and $\sim$300 mesh), while S3 (Xiaojiaao) into three samples (>150 mesh, 150–200 mesh, and <200 mesh). For oxygen and sulfur isotope analysis, the S1 and S2 samples were crashed and grounded into 200 meshes. The lead isotopes were measured with the MAT 261 mass spectrometer, the oxygen isotopes with the MAT 251, and the sulfur isotopes with the MAT 230 mass spectrometers (Lu and Wang 1990). The results of measurements are listed in Tables 1.3, 1.4, and 1.5, respectively.

It can be found from the data in Table 1.3 that the lead isotope composition of Suizhou is not uniform. It was just this ununiformity that gave rise a good Pb–Pb

Table 1.3 Results of lead isotope analysis for the Suizhou meteorite[a]

Sample	Mesh	Location	$^{206}Pb/^{204}Pb$	$^{207}Pb/^{204}Pb$	$^{208}Pb/^{204}Pb$
S_{1-1}	200–300	Dayanpo	18.745 ± 0.1	15.854 ± 0.04	28.893 ± 0.09
S_{1-2}	<300	Dayanpo	18.893 ± 0.003	15.907 ± 0.004	28.869 ± 0.001
S_{2-1}	100–300	Minjiachong	18.345 ± 0.01	15.677 ± 0.01	28.480 ± 0.03
S_{2-2}	<300	Minjiachong	18.563 ± 0.01	15.860 ± 0.004	28.957 ± 0.003
S_{3-1}	>150	Xiaojiaao	18.815 ± 0.03	15.980 ± 0.02	28.243 ± 0.09
S_{3-2}	150–200	Xiaojiaao	18.149 ± 0.005	15.633 ± 0.003	28.155 ± 0.004
S_{3-3}	<200	Xiaojiaao	18.157 ± 0.01	15.667 ± 0.001	28.261 ± 0.04

[a]Data from Lu and Wang (1990)

Table 1.4 Results of oxygen isotope analysis for the Suizhou meteorite[a]

Sample	Location	$\delta^{18}O$ (‰)/(SMOW)	Standard deviation
S_{1-1}	Dayanpo	+3.99	0.09
S_{1-2}	Dayanpo	+3.91	0.06
S_{1-3}	Dayanpo	+4.01	0.07
S_{2-1}	Minjiachong	+3.92	0.021
S_{2-2}	Minjiachong	+3.93	0.014
S_{2-3}	Minjiachong	+4.00	0.035

[a]Data from Lu and Wang (1990)

Table 1.5 Sulfur isotope values of the Suizhou meteorite[a]

Sample	Location	Sulfur content (wt %)	$\delta^{34}S$ (‰)
S_1	Dayanpo	2.312	+2.01
S_2	Minjiachong	2.371	+1.17

[a]Data from Lu and Wang (1990)

isochron, and from that isochron, the $^{207}Pb/^{204}Pb$–$^{206}Pb/^{204}Pb$ isochron age for the Suizhou meteorite was calculated as (4580 ± 0.6) Ma (Lu and Wang 1990). This age represents the formation age of parent body of the Suizhou meteorite.

Table 1.4 shows that the average values of $\delta^{18}O$ for S1 and S2 are 3.97 and 3.95 ‰, respectively, indicating that the distribution of oxygen isotope composition of the Suizhou meteorite is rather uniform. On the other hand, the sulfur contents for S1 and S2 are 2.312 and 2.371 %, respectively, but their $\delta^{34}S$ values (2.01 and 1.17 ‰) are different (Fig. 1.5), implying that a physicochemical process might be existed in this closed system.

1.5.2 Noble Gases in the Suizhou Meteorite

A fragment of the Suizhou meteorite taken from Weijiawan was used for noble gas analysis. Both the cosmogenic and trapped noble gases, such as He, Ne, Ar, Kr, and Xe, were measured by a mass spectrometer. The results of measurements and the calculated cosmic-ray exposure age and the K-^{40}Ar gas retaining age of the Suizhou meteorite are listed in Tables 1.6 and 1.7, respectively. It is found that the contents of trapped noble gases of ^{83}Kr and ^{126}Xe are consistent with the values for type 6 chondrites. Table 1.6 demonstrates that the average cosmic-ray exposure age of the Suizhou meteorite is (30.8 ± 3.3) Ma, indicating that the noble gases were very well retained in the Suizhou meteorite and the parent body of Suizhou meteor experienced a collision event at about 31 Ma ago which caused different shock features in minerals and formation of shock melt veins with abundant high-pressure phases in the Suizhou meteorite (Wang and Li 1990).

Table 1.7 also shows that the Suizhou meteorite has a K-^{40}Ar gas retaining age of (4280 ± 160) Ma. This age is younger than its formation age of 4580 Ma, indicating some radiogenic ^{40}Ar losses since the formation of the meteorite. However, there is no loss of cosmic nuclide ^{3}He since the Suizhou meteorite was exposed to cosmic ray soon after it was separated from its parent body. So the loss of radiogenic ^{40}Ar took place before the formation of the Suizhou meteor, i.e., at the stage of thermal metamorphism of the parent body or at the stage of the Suizhou meteor being separated from the parent body (Wang and Li 1990).

1.5.3 Fission Track Dating

The fission track dating and the thermal history of the Suizhou meteorite were studied by using the fission track technique on Suizhou pyroxenes (Zhang and Li 1990). The results show that the Suizhou pyroxenes have an average old track density of (7.0 ± 1.1) × 10^3/cm^2. According to this spontaneous track density and the track density induced by thermal neutrons of the reactor, the fission track age of Suizhou pyroxenes was calculated to be (3700 ± 300) Ma, indicating that pyroxenes in the parent body of the Suizhou meteorite began retaining fission tracks at about 3700 Ma ago. This age is obviously lower than the formation age (4580 Ma) and the K-^{40}Ar gas retaining age (4280 Ma) of the Suizhou parent body. This phenomenon may be related to the fact that the Suizhou parent body experienced remarkable thermal metamorphism after its formation, for the Suizhou meteorite is a type 6 chondrite.

Table 1.6 Analyses of noble gases in the Suizhou meteorite[a]

Cosmogenic noble gases								Trapped noble gases		
^{3}He	^{21}Ne	^{38}Ar	^{81}Kr	^{83}Kr	^{126}Xe	^{3}He/^{21}Ne	^{22}Ne/^{21}Ne	^{81}Kr	^{83}Kr	^{126}Xe
(10^{-8})	(cm^{3})	(STP/g)	(10^{-12})	$(cm^{3}\ STP/g)$				(10^{-10})	$(cm^{3}\ STP/g)$	
49.3	10.2	1.37	0.0166	3.71	0.248	4.833	1.084	116	0.20	0.84

[a]Data from Wang and Li (1990)

Table 1.7 Cosmic-ray exposure age and K-^{40}Ar gas retaining age of the Suizhou meteorite (Ma)[a]

Cosmic-ray exposure age							K-^{40}Ar gas retaining age
^{3}He	^{21}Ne	^{38}Ar	^{83}Kr	^{126}Xe	^{81}Kr-Kr	T$_{Avg.}$	
28.3	30.4	28.5	26.9	34.4	36 ± 8	30.8 ± 3.3	4280 ± 160

[a]Data from Wang and Li (1990)

1.6 Formation and Evolution of the Suizhou Meteorite

On-the-spot investigations together with an integrated study on the mineralogy, petrology, chemical composition, isotropic geochronology, fission track dating, noble gases, and stable isotopes of oxygen and sulfur have shed light on the formation and evolutional history of the Suizhou meteorite which is outlined by Li and Wang (1990) with a little modifications by authors of this book as follows:

1. Fractionation and condensation–agglomeration of the nebula and formation of mineral assemblages

The Suizhou parent body was formed about 4580 Ma ago as a result of the gathering and solidification of nebular condensates. According to the hypothesis of Sun–planet syngenesis and the theory of element-equilibrium condensation-agglomeration and the assemblages of major minerals identified in the Suizhou meteorite, the formation of minerals in this meteorite is thought to involve the following stages:

- At $\geqq$1600 K, the refractory minerals condensed to form ilmenite, chromite, and, might be, graphite.
- At 1600–1400 K, the FeNi metal condensed to form kamacite and taenite.
- At 1500–1200 K, silicates of Ca, Mg, and Fe condensed to form olivine, orthopyroxene, and clinopyroxene.
- At 1100–1000 K, silicates and phosphates of alkali metals condensed to form plagioclase, whitlockite, and apatite.
- At 1000 to −570 K, sulfides condensed to form troilite and other sulfide minerals.

In the Suizhou meteorite, the major refractory condensates, chromite and ilmenite, are largely distributed in chondrules, while the principal low-temperature condensate, troilite, is distributed mainly in matrix, indicating that the chondrules were formed earlier than the matrix.

2. The thermal history of the Suizhou parent body

After the formation of the Suizhou parent body as a result of condensation–agglomeration and accretion of nonvolatile condensates in the nebula, the parent body experienced thermal metamorphism under the action of solar wind, accretion of gravitational energy, and radioactive energy. As estimated from the order parameter of carbonaceous material, the temperature of thermal metamorphism is ≤ 777 °C in the position where the Suizhou meteorite was located. The principal thermal metamorphic marks are presented as follows:

- Compositional homogenization of olivine and pyroxene and increase of mineral grain size.
- The ordering degree of pyroxene is increased as indicated by the change in Fe^{2+} occupancy from disordered distribution (occupying the Mi and M2 sites) for nonequilibrium chondrites to completely ordered distribution (only occupying the M2 site) for the Suizhou meteorite.
- Indistinct outlines of most chondrules.
- Ordering and graphitization of carbonaceous material.
- Devitrification and recrystallization of glassy materials.
- The concentration of trapped noble gases and volatile elements is much lower than those in the low petrologic types of chondrites.

At 4280 Ma ago, the Suizhou parent body cooled to the temperature at which argon (Ar) might be retained, and at about 3700 Ma ago, the pyroxenes in the Suizhou parent body began retaining fission tracks. Since the thermal metamorphism in the parent body is a factor controlling the petrochemical type of a chondrite, the above-mentioned thermal metamorphic marks are indicative for that the Suizhou meteorite is a highly thermal metamorphosed meteorite, and it should be assigned to L6 type.

3. Formation of the Suizhou meteorite

As determined on the basis of the cosmogenic nuclides and ^{81}Kr-Kr dating, the Suizhou meteorite has an average cosmic-ray exposure age of (30.8 ± 3.3) Ma. The results obtained by different methods are of good consistency within the limit of experimental errors, indicating that the Suizhou meteorite has only a single-stage exposure history. The event that the parent body of the Suizhou meteorite broke up as a result of collision to form the Suizhou meteor occurred about 30.8 Ma ago. It was also revealed that the loss of radiogenic ^{40}Ar took place before the formation of the Suizhou meteor.

As a result of the collision event, the Suizhou meteor experienced strong shock metamorphism, and abundant of various high-pressure mineral phases was formed in the thin shock melt veins during this collision event.

4. Falling of Suizhou meteorite and formation of meteorite shower

On April 15, 1986, at 18:52 Beijing time, the Suizhou meteor entered the atmosphere at a relative steep angle over Dayanpo near the Suizhou City and violently exploded at the altitude of about 10,000 m, followed by the secondary explosion—fragmentation at lower altitude.

During its atmospheric passage, the Suizhou meteor suffered ablation under high-temperature and low-pressure conditions. As a result, the surfaces of the meteor fragments were fused, forming the fusion crust of 0.5–1.5 in thickness.

The falling area of the Suizhou meteorite is about 10 km long from east to west and 5 km wide from south to north, totaling about 39 km^2. The recovered meteorite fragments weigh $\geqq$210 kg. As estimated from the preliminary determinations of cosmogenic nuclides He and Ne, the total mass of the Suizhou meteorite weigh $\geqq$270 kg, the second large stone meteorite and stone meteorite shower (after the Jilin meteorite) so far known in China.

References

Chen JZ (1990) A transmission electron microscopic study of the Suizhou meteorite. A synthetical study of Suizhou meteorite. Publishing House of the China University of Geosciences, Wuhan, pp 57–62 (in Chinese)

Chen M, Shu JF, Xie XD et al (2003a) Natural CaTi$_2$O$_4$-structured FeCr$_2$O$_4$ polymorph in the Suizhou meteorite and its significance in mantle mineralogy. Geochim Cosmochim Acta 67 (20):3937–3942

Chen M, Shu J, Mao HK et al (2003b) Natural occurrence and synthesis of two new postspinel polymorphs of chromite. Proc Natl Acad Sci USA 100(25):14651–14654

Hou W, Ouyang ZY, Li ZH et al (1990) Suizhou, the new fall of a L-group chondrite. In: A synthetical study of Suizhou meteorite. Publishing House of the China University of Geosciences, Wuhan, pp 30–32 (in Chinese)

Huang DG, Xiao HY (1990) Bulk chemical analyses of Suizhou meteorite. In: A synthetical study of Suizhou meteorite. Publishing House of the China University of Geosciences, Wuhan, pp 80–82 (in Chinese)

Li ZH, Wang RJ (1990) Formation and evolution of the Suizhou meteorite. In: A synthetical study of Suizhou meteorite. Publishing House of the China University of Geosciences, Wuhan, pp 131–143 (in Chinese)

Li ZH, Wang DD, Ouyang ZY et al (1990) A raman spectroscopic study of the Suizhou metorite. In: A synthetical study of Suizhou meteorite. Publishing House of the China University of Geosciences, Wuhan, pp 91–96 (in Chinese)

Lu YL, Wang RJ (1990) A lead, oxygen and sulfur isotope study of the Suizhou meteorite. In: A synthetical study of Suizhou meteorite. Publishing House of the China University of Geosciences, Wuhan, pp 120–125 (in Chinese)

Qiu JR, Wang RJ, Pan DJ (1990) The fusion crust constituent of the Suizhou meteorite and its significance. In: A synthetical study of Suizhou meteorite. Publishing House of the China University of Geosciences, Wuhan, pp 19–30 (in Chinese)

Shen JC (1990) Two ixiolite-structured new Nb-Ta minerals found in the Suizhou meteorite. In: A synthetical study of Suizhou meteorite. Publishing House of the China University of Geosciences, Wuhan, pp 52–54 (in Chinese)

Shen JC, Ren YX (1990) Identification of fine minerals in Suizhou meteorite by single crystal X-ray diffraction method. In: A synthetical study of Suizhou meteorite. Publishing House of the China University of Geosciences, Wuhan, pp 55–56 (in Chinese)

Shen SY, Zhuang XL (1990) A study of the opaque minerals and structural characteristics of the Suizhou meteorite. In: A synthetical study of Suizhou meteorite. Publishing House of the China University of Geosciences, Wuhan, pp 40–52 (in Chinese)

Shu QR, Yuan YM (1990) Falling phenomena of the Suizhou meteorite. In: A synthetical study of Suizhou meteorite. Publishing House of the China University of Geosciences, Wuhan, pp 1–7 (in Chinese)

Wang DD (1990) Characteristics of the trace element abundance of the Suizhou meteorite. In: A synthetical study of Suizhou meteorite. Publishing House of the China University of Geosciences, Wuhan, pp 77–79 (in Chinese)

Wang DD, Li ZH (1990) A study of rare gas dating of the Suizhou meteorite. In: A synthetical study of Suizhou meteorite. Publishing House of the China University of Geosciences, Wuhan, pp 125–128 (in Chinese)

Wang RJ, Qiu JR, Shen JC (1990a) A preliminary study of the Suizhou meteorite. In: A synthetical study of Suizhou meteorite. Publishing House of the China University of Geosciences, Wuhan, pp 12–19 (in Chinese)

Wang RJ, Qiu JR, Shen SY (1990b) Chondrules of the Suizhou meteorite and their formation. In: A synthetical study of Suizhou meteorite. Publishing House of the China University of Geosciences, Wuhan, pp 65–72 (in Chinese)

Xie XD, Chen M, Wang DQ (2001a) Shock-related mineralogical features and P-T history of the Suizhou L6 chondrite. Eur J Miner 13(6):1177–1190

Xie XD, Chen M, Wang DQ et al (2001b) $NaAlSi_3O_8$-hollandite and other high-pressure minerals in the shock melt veins of the Suizhou L6 chondrite. Chin Sci Bull 46(13):1121–1126

Xie XD, Chen M, Dai CD et al (2001c) A comparative study of naturally and experimentally shocked chondrites. Earth Planet Sci Lett 187:345–356

Xie XD, Minitti ME, Chen M et al (2002) Natural high-pressure polymorph of merrillite in the shock vein of the Suizhou meteorite. Geochim Cosmochim Acta 66:2439–2444

Xie XD, Minitti ME, Chen M et al (2003) Tuite, γ-Ca3(PO4)2, a new phosphate mineral from the Suizhou L6 chondrite. Eur J Miner 15:1001–1005

Xie XD, Sun ZY, Chen M (2011a) The distinct morphological and petrological features of shock melt veins in the Suizhou L6 chondrite. Meteor Planet Sci 46:459–469

Xie XD, Chen M, Wang CW (2011b) Occurrence and mineral chemistry of chromite and xieite in the Suizhou L6 chondrite. Sci China Earth Sci 54:1–13

Zeng GC (1990) Common transparent minerals and chemico-petrological type of the Suizhou meteorite. In: A synthetical study of Suizhou meteorite. Publishing House of the China University of Geosciences, Wuhan, pp 32–40 (in Chinese)

Zhang F, Li ZH (1990) Fission track dating of the Suizhou meteorite. In: A synthetical study of Suizhou meteorite. Publishing House of the China University of Geosciences, Wuhan, pp 128–130 (in Chinese)

Zhong HH, Jiang LJ, Yang XH et al (1990) A preliminary study of the micro-elements in the Suizhou meteorite. In: A synthetical studiy of Suizhou meteorite. Publishing House of the China University of Geosciences, Wuhan, pp 74–77 (in Chinese)

Chapter 2
Micro-Mineralogical Investigative Techniques

This chapter gives a summary of techniques used in the investigations of minerals, particularly the fine-grained minerals in the Suizhou meteorite. For the various techniques summarized by Edward C.T. Chao and Xiande Xie in their two publications (Chao and Xie 1989, 1990), here we try to describe techniques needed and suitable for the investigation of shock-induced fine structures within minerals, as well as for identification and characterization of tiny new high-pressure polymorphs produced by shock, including optical microscopy, scanning electron microscopy (SEM) energy-dispersive X-ray analysis (EDS), electron microprobe (EPMA), Raman microprobe analysis (RMA), synchrotron radiation X-ray diffraction (SRXD), X-ray micro-diffraction (XRMD), transmission electron microscopy (TEM), and laser ablation inductively coupled plasma mass spectrometry (LA-ICP-MS) to provide examples of optimal use of the various techniques and how they complement each other.

2.1 Petrographic Microscopic Studies

The petrographic microscope is a basic tool for the use to identify minerals and petrologic type of meteorite, mineral assemblages, and structural and textural relationships and thus infer paragenesis. Petrographic microscopic study is a preliminary stage of meteorite research needed to pinpoint identification of minerals and selection of samples for detailed SEM, EDS, and micro-Raman analyses, as well as to guide other aspects of multidisciplinary investigations. One of the principal objectives of using polished thin sections (PTSs) of the Suizhou meteorite was to study the internal structures and deformation features of an impact process in minerals. Such deformational process produces effects that reflect fast strain rates generated by meteorite impact. Microscopic investigations are of great importance in the study of shock metamorphic features of olivine, pyroxene, and plagioclase in a meteorite and thus classify its shock degree. Another objective is to find shock-induced high-pressure phases in shock melt veins. Microscopic study is also important for photographical documentation of the locations of the minerals in PTS for further SEM-EDS, EPMA, and RMA.

© Springer-Verlag Berlin Heidelberg and Guangdong Science
& Technology Press Co., Ltd. 2016
X. Xie and M. Chen, *Suizhou Meteorite: Mineralogy and Shock Metamorphism*,
Springer Geochemistry/Mineralogy, DOI 10.1007/978-3-662-48479-1_2

Fig. 2.1 An MPV-SP optical
microscope

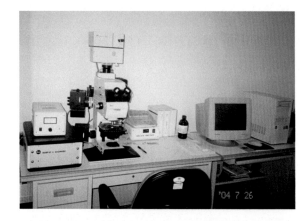

Fig. 2.2 Micrograph
showing shock-induced
fragmentation in olivine.
Crossed nicols

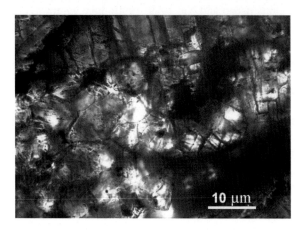

An MPV-SP optical microscope with both transmitted and reflected lights has
been used at Guangzhou Institute of Geochemistry, Chinese Academy of Sciences,
for petrographic microscopic studies of both transparent and opaque minerals in the
Suizhou meteorite (Fig. 2.1). The example given here is concerning the study of
shock-induced deformation features in olivine of the Suizhou meteorite (Fig. 2.2).

2.2   Scanning Electron Microscopy Energy-Dispersive
X-Ray Analysis

A scanning electron microscope equipped with appropriate detectors and elec-
tronics for EDX provides a capability for rapid semi-quantitative compositional
analysis of individual mineral and inclusions within grains. For textural analysis,
SEM extends the capability of the optical microscope with respect to depth of field

and spatial resolution (about 2.5 nm). The three-dimensional aspect of the images obtained by SEM is of particular value for applications such as characterization of crystal morphology, porosity studies, and deformation features. Images obtained using back-scattered electrons (BSEs) show variations in intensity which highlight differences in average atomic number, thus revealing the different mineral phases present and serving as a guide for more efficient SEM-EDX and EPMA analysis.

For SEM-EDX studies of the Suizhou meteorite, a Hitachi S-3500 N scanning electron microscope equipped with a Link ISIS 300 X-ray energy-dispersive spectrometer has been used at Guangzhou Institute of Geochemistry (Fig. 2.3). In analyses, the minimum detection level for most elements for routine SEM-EDX analysis is approximately 0.1–0.2 wt%. In addition, our EDX can detect elements present in particles as small as about 1 μm. Figure 2.4 is an SEM image in back-scattered electron (BSE) mode showing a shock-induced melt vein in the Suizhou meteorite and different minerals outside the vein.

Fig. 2.3 The Hitachi S-3500 N scanning electron microscope with a Link ISIS 300 X-ray energy-dispersive spectrometer

Fig. 2.4 Back-scattered electron image showing a shock vein in the Suizhou meteorite (Xie et al. 2011)

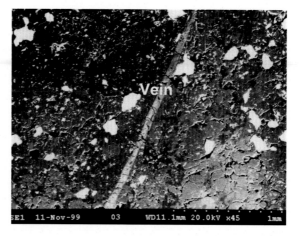

2.3   Electron Probe Microanalysis

Quantitative analysis with the electron microprobe requires more elaborate sample preparation than for most SEM studies: Surface for analysis must be flat, polished, and as scratch free as possible. Thus, specimens for EPMA must be PTS or polished mounts. The outstanding advantage of EPMA is the ability to carry out in-place identification of a minute mineral or crystal in a prepared sample without having to extract the grain for study.

A Cameca SX-51 electron microprobe at the Institute of Geology and Geophysics, Chinese Academy of Sciences, and a JEOL JXA-8100 electron microprobe at the Guangzhou Institute of Geochemistry, Chinese Academy of Sciences (Fig. 2.5), were used for analyzing the Suizhou minerals. Minimum detection limits for our EPMA for routine analysis of most elements are of the order of 500 ppm, and the minimum area of analysis is a circle of about 3 μm in diameter. Depth of penetration of the incident electron beam into the target is about 3 μm in most Suizhou minerals. The capability for quantitative analysis of such small areas makes EPMA invaluable to us for (1) mineral identification of the Suizhou meteorite on the basis of concentration of major and minor elements present, especially in cases where the sensitivity of analysis by SEM-EDX may not be sufficient for identification (Fig. 2.6); (2) identification of phases transformed in solid state or crystallized and quenched from shock-induced melt at high pressures and temperatures; (3) determination of the chemical formula of Suizhou minerals and high-pressure polymorphs and assignments of cations to site occupancies, as in analyzing new minerals tuite and $CaFe_2O_4$- and $CaTi_2O_4$-structured polymorphs of chromite; (4) mapping of elements in different phases in the shock melt veins; and (5) study of melt-vein minerals with incorporation of additional elements from the surrounding matrix melt.

Fig. 2.5 The JEOL JXA-8100 microprobe

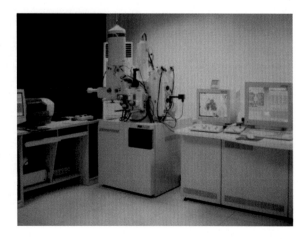

Fig. 2.6 BSE image of an
apatite grain (Apt)

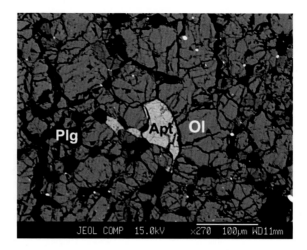

2.4 Raman Microprobe Analysis

With the recent development of micro-Raman techniques, the Raman microprobe is emerging as an important tool in mineralogical research. Raman spectroscopy provides detailed structural information about site symmetry, short-range and long-range bonding, and lattice vibrational properties. Macro-sampling Raman spectroscopy has been used as the main analytical method of studying the structures of silicate glasses quenched from various temperatures and pressures. The method provides critical structural information such as the mechanism of pressure densification, the angle of Si–O–Si linkages, the number of nonbridging oxygens, and the effects of ion substitution (Chao and Xie 1989, 1990; Fukukawa 1981; McMillan 1984; Mysen et al. 1982). Raman spectra are also sensitive to slight differences in the symmetry of polymorphous mineral phases and to the ordering of cations.

The coupling of a research-grade microscope to a Raman spectrometer has led to the capability of obtaining Raman spectra for micron-sized samples. A Raman microprobe can be used as a routine petrographic tool comparable to a conventional petrographic microscope. High spatial resolution and rapid measurement can be obtained with regular PTS. In addition to "fingerprinting" identification of micron-sized minerals, RMA yields sophisticated structural information of shock-metamorphosed minerals. The broadening of a Raman band in a shocked mineral may be indicative of pressure densification and changing in bonding angles. The additional band in the Raman spectrum of a shocked mineral may reflect shock-produced disorder. The changing and shifting of Raman bands may reflect the structural transformation of a shocked mineral. Therefore, obtaining this crystal chemical information for minerals in place in thin section is critical to understanding the history and the evolution process of a meteorite.

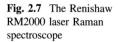

Fig. 2.7 The Renishaw
RM2000 laser Raman
spectroscope

A Renishaw RM2000 laser Raman microprobe (Ar^+ laser, 514.5-nm line) in the
Guangzhou Institute of Geochemistry, Chinese Academy of Sciences (Fig. 2.7), and
the RM1000 laser Raman microprobes (Ar^+ laser, 514.5-nm line) in the Beijing
Institute of Nonferrous Metals as well as in the China University of Geosciences
have been used for recording Raman spectra of different minerals and polymorphous phases of the Suizhou meteorite. An example of RMA application is the
study of shock metamorphic features of plagioclase in Suizhou. The Raman
spectrum of the new mineral tuite, the shock-induced high-pressure polymorph of
whitlockite, in the Suizhou melt vein is shown in Fig. 2.8.

2.5   Synchrotron Radiation X-Ray Diffraction in Situ Analysis

Synchrotron radiation energy-dispersive X-ray diffraction (SRXRD) technique
developed in the Geophysical Laboratory of the Carnegie Institution of Washington
is one of the best techniques for in situ studying of micron-sized polycrystalline
grains or single crystals under high pressures or at ambient conditions, as well as for
studying inclusions and lamellae in minerals (Hemley 1998). This technique uses
the entire energy spectrum of synchrotron radiation. The collimated polychromatic
(white) X-ray beam impinges on the mineral specimen. The polychromatic diffracted beam is collimated at a fixed 2θ angle and collected by a solid-state detector
which disperses the diffracted photons in the energy spectrum, and at a fixed Bragg
angle 2θ, and the *d*-spacings are determined from the peak energies (Hemley 1998).
Minute samples and samples of low diffraction intensity (low atomic number) can
be studied with this technique.

SRXRD has been used for studying the fine-grained minerals and polymorphs of
the Suizhou meteorite. After investigation with optical microscope, scanning SEM,
EPMA, and RMA, the micron-sized target mineral grains in thin sections were

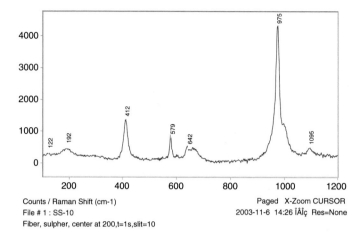

Fig. 2.8 The Raman spectrum of the high-pressure polymorph of whitlockite in Suizhou

analyzed in situ by energy-dispersive X-ray diffraction using the synchrotron beamline X17C at the National Synchrotron Light Source, Brookhaven National Laboratory, USA (Fig. 2.9). The operating voltage was 2.584 GeV and the current —300 to 100 mA. X-ray beam was collimated to a size of μm 15 × 15 and was focused on the probed mineral sample in the PTS which was rotated systematically ($\omega = -30°-30°$, $\chi = 0°-360°$) to collect diffraction lines. Energy-dispersive X-ray diffraction (EDXD) was gathered with an intrinsic germanium detector. The X-ray diffraction data were acquired at 2θ settings of 8° and 10°, respectively.

It should be pointed out that using the advanced SRXRD technique, the authors of this book were able to obtain the X-ray diffraction patterns for the fine-grained vein matrix minerals and three shock-produced and micron-sized new minerals, namely tuite, the high-pressure polymorph of whitlockite, and two post-spinel high-pressure phases, the $CaFe_2O_4$-type and $CaTi_2O_4$-type polymorphs of chromite in the Suizhou meteorite, and the three synthetic quenched products recovered from high *P-T* experiments were also analyzed by synchrotron X-ray diffraction at the

Fig. 2.9 The synchrotron radiation X-ray diffraction device at Brookhaven National Laboratory

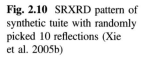

Fig. 2.10 SRXRD pattern of synthetic tuite with randomly picked 10 reflections (Xie et al. 2005b)

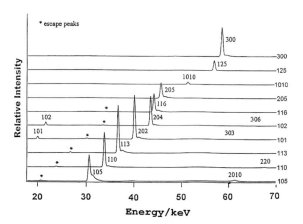

X17C superconducting wiggler beamline of the National Synchrotron Light Source, Brookhaven National Laboratory, USA, for phase identification and crystal structure determination (Xie et al. 2005; Chen et al. 2003a, b) (Fig. 2.10).

The utility of synchrotron radiation techniques has exceeded the development of high-pressure mineralogy. At the same time, the techniques developed for in situ studies at high pressures and temperatures are being used to investigate microscopic inclusions such as coesite in high-pressure metamorphic rocks and deep mantle samples as inclusions in diamond (Hemley 1998). The results of our investigations added some more examples of using synchrotron radiation technique for in situ investigations of fine-grained mineral aggregates and identification of extremely small-sized new minerals embedded in the fine-grained matrix.

2.6 X-Ray Micro-Diffraction in Situ Analysis

X-ray micro-diffractometer is an effective tool for obtaining X-ray diffraction data of tiny size of minerals and materials, usually less than 1 mm, directly on thin sections or plane blocks containing the mineral or material of interest. The structure of the instrument is similar to four-circle single-crystal diffractometer, but differs in lacking the χ-axis in the former. The RIGAKU D/Max Rapid IIR X-ray micro-diffractometer in the Central South University has been used in in situ study of some Suizhou minerals. It is mainly composed of the following parts (Fig. 2.11): (1) X-ray generator with the maximum power of 18 KW; (2) X-ray collimator tube of diameters 0.1, 0.05, 0.03, and 0.01 mm; (3) two-axis goniometer defined as the φ-axis (vertical to the sample plane) and ω-axis (vertical to the horizontal plane); (4) X-photon detector of high-sensitivity two-dimensional image plate or CCD setup as a cylinder with the diameter around 127.4 mm; (5) monitoring video for sample positioning with magnitude from 30 to 240; and (6) controlling software.

Fig. 2.11 The RIGAKU
D/Max Rapid IIR X-ray
micro-diffractometer in situ
analysis

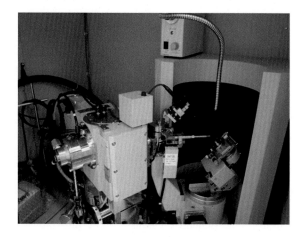

The diffraction effect of the sample is recorded on 2D image plate of the area (470×256) mm^2 arranged in 4700×2560 pixels. The pixel coordinates of a diffraction dot on the image are related to the incident angle (θ) of X-ray and the dipping angle (β) of the normal line of the sample plane. Numerous diffraction dots with the same θ constitute a Debye ring. Intensity integration along the Debye rings yields one-dimensional 2θ-I data similar to the pattern of powder diffraction (Fig. 2.12).

The area of the sample covered by X-ray depends upon the diameter (d) of collimator, and the angle ω between incident X-ray and the sample plane and the starting 2θ in the diffraction data is close to ω. In practice, the minimum grain size required for a pure phase diffraction data is around 0.15 mm for 0.03-mm collimator and $\omega = 20°$, and around 0.25 mm for 0.05-mm collimator and $\omega = 20°$.

Various forms of samples may be accepted for X-ray micro-diffractometer, including polished thin section or polished block, powder, small specimen, and a single tiny grain. The minimum grain size of sample to have distinguishable diffraction lines (dots) varies with the diffraction ability expressed by K value and

Fig. 2.12 The X-ray
diffraction pattern taken by
micro-diffractometer

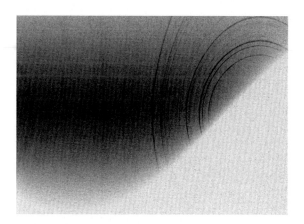

the crystallinity as well as the iron contents, to say, as small as 2 microns for minerals of high K values, such as galena, or as big as 40 microns for clay minerals of low K values, such as kaolinite. The exposure time to get good results varies with the collimator diameter and the grain size of the sample, to be as short as 10 min for 0.1-mm collimator on a grain size of 0.5 mm, or to be as long as 12 h for 0.03-mm collimator on a grain size of 0.05 mm.

2.7   Transmission Electron Microscopy

Microstructures or domains in minerals are indicative of the geological environment and the processes involved in their formation and evolution. Microstructures in rock-forming minerals resulting from pressure, temperature, and composition conditions during crystallization and cooling include atomic site ordering, exsolution, and phase transitions. Fine internal structures produced by deformational processes may be characteristic of particular ranges of strain rates. However, microstructures are too small to be resolved either by optical methods or by SEM, but they can be studied using TEM. Prior to the development of thinning by ionic bombardment, samples thin enough to be transparent to electron beams could generally be produced only as dispersed small particles or flakes. Chemically thinned individual grains or surface replicas and ultramicrotome sections are prepared from bulk specimens. It was only after the advent of ion thinning that it became possible to heavily utilize TEM in geological and mineralogical studies. Another aspect using TEM is the acquisition of electron diffraction data, either SAED (selected area electron diffraction) or general area diffraction. This allows one to identify very small mineral grains while studying its substructures. TEM is particularly suitable for identification of inclusions or high-pressure polymorphs less than 5 μm in size which are difficult to extract in order to obtain XRD data, or identification of submicron-sized minerals observed optically.

A Hitachi transmission electron microscope and a JEM 2100F field-emission transmission electron microscope (Fig. 2.13) at the Wuhan University of Technology were used for studying the mineralogy and fine structures of minerals of the Suizhou meteorite. Figure 2.14 is a TEM micrograph showing the shock-induced fine-grained high-pressure polymorphs in a Suizhou melt vein.

Very recently, the focused ion beam (FIB) method has been developed for fast preparing of the TEM samples in material sciences (Zhou and Xu 2004). This technique can be used to cut thin TEM samples in different fixed locations on PTS of terrestrial or extraterrestrial rocks. One of such kinds of instruments is Seiko SMI 2200 focused ion beam system produced by the Seiko Company of Japan. There are two methods of preparing TEM samples. One is the traditional trench method. At the first stage, a specimen is cut to a 3 mm × 0.1 mm size and ion-thinned to a thickness of 20–30 μm. Then, a TEM sample of 10 μm × 5 μm with a thickness

Fig. 2.13 JEM 2100F field-emission transmission electron microscope

Fig. 2.14 TEM micrograph showing fine-grained high-pressure polymorphs in a Suizhou melt vein

of 0.1 μm is prepared by using the FIB system. The other is the lift-out method. At first, a carbon layer is deposited on the sample to protect the sample surface. Then, a cross section will be engraved on one side of the sample on polished thin section, and afterward, another cross section also be cut on the other side of the sample. However, the thickness of the sample is about 0.8 μm that is thicker than that required for TEM study. For further thinning, the sample holder should be tilted to 45° for cutting through the sample's lower edge and then return the sample holder to its original position. After the sample being thinned to 0.2 μm in thickness, cut through the foil sample at two tops; thus, a suitable TEM sample is prepared. Finally, lift out the prepared sample and lay it on a carbon-coated copper grid by using a specimen transfer.

2.8 Laser Ablation ICP-MS

The laser ablation inductively coupled plasma mass spectrometry (LA-ICP-MS) can precisely determine trace element concentrations of geological samples ranging from ultramafic to granitic and the precision and accuracy for elements with concentrations higher than 0.1 µg/g (Tu et al. 2011). In LA-ICP-MS, the sample is directly analyzed by ablating with a pulsed laser beam. The created aerosols are then transported into the core of inductively coupled argon plasma (ICP), which generates temperature of approximately 8000 °C. The plasma in ICP-MS is used to generate ions that are then introduced to the mass analyzer. These ions are then separated and collected according to their mass-to-charge ratios. The constituents of an unknown sample can then be identified and measured. ICP-MS offers extremely high sensitivity to a wide range of elements. For laser ablation, any type of solid sample can be ablated for analysis; there are no sample-size requirements and no sample preparation procedures.

Chemical analysis using laser ablation requires a smaller amount of sample (micrograms) than that required for solution nebulization (milligrams). Depending on the analytical measurement system, very small amount of sample quantities may be sufficient for this technique. In addition, focused laser beam permits spatial characterization of heterogeneity in solid samples, with typically micron resolution in terms of both lateral and depth conditions.

In the study of our Suizhou minerals, an Agilent 7500a coupled with a Resonetics RESOlution M-50 laser ablation system in the Guangzhou Institute of Geochemistry, Chinese Academy of Sciences, was used to analyze about 40 trace elements for all minerals (Fig. 2.15). It consists of an excimer (193 nm) laser, a two-volume laser ablation cell, a squid smoothing device, and a computer-controlled high-precision X-Y stage. The two-volume laser ablation cell is designed to avoid cross-contamination and reduce background flushing time. The squid smoothing device can reduce statistic error induced by laser ablation pulses.

Fig. 2.15 The Agilent 7500a ICP-MS coupled with a Resonetics RESOLution M-50 laser ablation system

Fig. 2.16 BSE image showing a black laser ablation hole (*arrow*) in the Suizhou majorite grain

The accuracy of the X-Y stage is better than 0.1 μm. Laser ablation was operated at a constant energy 60 mJ and a repetition rate of 8 Hz, with a spot diameter of 31 μm (Fig. 2.16), using external standards of NIST SRM 610, NIST SRM 661, and GOR 128 from Ding-Glass (Pearce et al. 1997; Jochum et al. 2005). The primary data used two different internal standards to calibration of ^{29}Si in silica melt and ^{59}Ni as the FeNi metal grains. The relative standard deviations are mostly less than 5 %, and relative standard deviations of obtained average concentrations from reference values are mostly less than 10 % (Tu et al. 2011). Data reduction was carried out using ICPMSDataCal (Liu et al. 2008).

References

Chao ECT, Xie XD (1989) Micromineralogical techniques to geological investigations. Science Press, Beijing, p 215 (in Chinese)

Chao ECT, Xie XD (1990) Mineralogical approaches to geological investigations. Science Press, Beijing, p 388

Chen M, Shu JF, Xie XD et al (2003a) Natural CaTi$_2$O$_4$-structured FeCr$_2$O$_4$ polymorph in the Suizhou meteorite and its significance in mantle mineralogy. Geochim Cosmochim Acta 67 (20):3937–3942

Chen M, Shu J, Mao HK et al (2003b) Natural occurrence and synthesis of two new postspinel polymorphs of chromite. Proc Natl Acad Sci USA 100(25):14651–14654

Fukukawa T, Fox KE, White WB (1981) Raman spectroscopy investigation of the structure of silicate glasses. III. Raman intensities and structural units in sodium silicate glasses. J Chem Phys 75:3226–3237

Hemley RJ (1998) Preface, In: Ultrahigh-pressure mineralogy: physics and chemistry of the Earth's deep interior, reviews in mineralogy, vol 37, pp 3–8

Jochum P, Pfander J, Woodhead JD, Willbold M, Stoll B, Herwig K, Amini M, Abouchami W, Hofmann AW (2005) MPI-DING glasses: New geological reference materials for in situ Pb isotope analysis. Geochem Geophys Geosystems 6:1–44

Liu YS, Hu ZC, Gao S et al (2008) In situ analysis of major and trace elements of anhydrous minerals by LA-ICP-MS without applying an internal standard. Chem Geol 257(1–2):34–43

McMillan P (1984) Structural studies of silicate glasses and melts; application and limitations of Raman spectroscopy? Am Mineralogists 69:622–644

Mysen BO, Virgo D, Seifert FA (1982) The structure of silicate melts: implications for chemical and physical properties of natural magma. Rev Geophys 20:353–383

Pearce NJG, Perkins WT, Westgate JA et al (1997) A compilation of new and published major and trace element data for NIST SRM 610 and NIST SRM 612 glass reference materials. Geostand Newsl-J Geostand Geoanal 21(1):115–144

Tu XL, Zhang H, Deng WF et al (2011) Application of resolution in-situ laser ablation ICP-MS in trace element analyses. Geochemica 40(1):83–98 (in Chinese with English abstract)

Xie XD, Chen M, Wang DQ (2005a) Two types of silicate melts in naturally shocked meteorites. In: Papers and abstracts of the 5th annual meeting of IPACES, Guangzhou, pp 12–14

Xie XD, Shu JF, Chen M (2005b) Synchrotron radiation X-ray diffraction in-situ study of fine-grained minerals in shock veins of Suizhou meteorite. Sci China Ser D 48:815–821

Xie XD, Sun ZY, Chen M (2011) The distinct morphological and petrological features of shock melt veins in the Suizhou L6 chondrite. Meteorit Planet Sci 46:459–469

Zhou WM, Xu NH (2004) Rapid preparation of TEM samples by focused ion beam. J Chin Electr Microsc 23(4):513 (in Chinese)

Chapter 3
Mineralogy of Suizhou Unmelted Chondritic Rock

3.1 General Remarks

Type 6 ordinary chondrites generally have a simple mineralogy. Major minerals are olivine, low-Ca pyroxene, and FeNi metal (kamacite and taenite). Plagioclase, diopside, and troilite are minor minerals, and accessory minerals include chromite, whitlockite (merrillite), and chlorapatite. Native copper, pigeonite, pentlandite, ilmenite, mackinawite, bravoite, and chalcopyrite may also occur.

The Suizhou meteorite is one of the type 6 chondrites. The unmelted chondritic portion makes up more than 98 % of the Suizhou meteorite by volume. The remaining <2 % by volume of the meteorite consists of shock-induced melt veins formed by collision event in the space and fusion crust formed during its atmospheric passage. The minerals in the Suizhou unmelted chondritic portion include olivine, low-Ca pyroxene, plagioclase, maskelynite, kamacite, taenite, troilite, whitlockite, chlorapatite, chromite, ilmenite, and natural copper. The general introduction of the Suizhou rock-forming minerals as well as the minor and accessory minerals has been given in Chap. 1. In this chapter, we try to present our newly obtained results in the study of rock-forming minerals, opaque minerals, and accessory phosphate minerals with the emphasis on their shock-induced mineralogical features.

3.2 Rock-Forming Minerals

The rock-forming minerals of the Suizhou meteorite are olivine, pyroxene, and plagioclase. They consist of about 88 % of the meteorite by volume (Fig. 3.1). Study of their characteristics and shock-induced deformation features is of great importance in understanding of the formation and evolutionary history of the Suizhou meteorite. Maskelynite, a quenched dense plagioclase melt glass, also

© Springer-Verlag Berlin Heidelberg and Guangdong Science
& Technology Press Co., Ltd. 2016
X. Xie and M. Chen, *Suizhou Meteorite: Mineralogy and Shock Metamorphism*,
Springer Geochemistry/Mineralogy, DOI 10.1007/978-3-662-48479-1_3

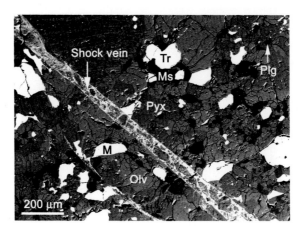

Fig. 3.1 A back-scattered electron (BSE) image showing a shock vein intersecting the unmelted chondritic portion of the Suizhou meteorite. Note the olivine (*Olv*), pyroxene (*Pyx*), and plagioclase (*Plg*) contain abundant fractures, while maskelynite (*Ms*) displays little fracturing in the interior. M = FeNi metal; Tr = troilite (Chen et al. 2004a)

occurs in the Suizhou unmelted chondritic rock. Its occurrence is also of some significance in the understanding the shock history of the Suizhou meteorite. Here, we introduce the physical and chemical properties of these four rock-forming minerals as well as their shock effects.

3.2.1   Olivine

Olivine is one of the most important rock-forming silicate minerals in the Earth's upper mantle, along with pyroxenes and garnets, and is a most common silicate mineral in stone meteorites. The geologically relevant end-member phases are forsterite (Fo), Mg_2SiO_4, and fayalite (Fa), Fe_2SiO_4, and the intermediate isomorphous mixture is olivine which has the composition of $(Mg,Fe)_2SiO_4$, with the natural composition being somewhat near $Fo_{90}Fa_{10}$. Fe^{2+} sometimes is replaced by Ca^{2+} or Mn^{2+} in the Fe-rich members, and Mg^{2+} is replaced by Cr^{3+} or Ni^{2+} in the Mg-rich members. The rare mineral tephroite (Te), Mn_2SiO_4, also belongs to the olivine group. The double salt, monticellite (Mo), $(Ca,Mg)_2SiO_4$, is closely related to olivine and is often considered to be a member of the group. The crystal system of olivine group members is orthorhombic. Larnite, with the composition Ca_2SiO_4, is not a member of the olivine group since it is monoclinic.

The structure of forsterite is based on a distorted hexagonal closely packed array of oxygen atoms with one tetrahedral site (T) and two nonequivalent octahedral sites (M_1, M_2). The M_1 site is more distorted than M_2 site largely because the M_1 octahedron shares 6 of its edges with neighboring polyhedra (2 with M_1, 2 with M_2, and 2 with T), while the M_2 octahedron only shares 3 edges (2 with M_1, and 1 with

T). According to Pauling's rules, shared edges are shorter than unshared edges in order to screen cation–cation repulsion (Prewitt and Downs 1998).

The shock effects in olivine in ordinary chondrites were studied on the basis of optical microscopy by Stöffler et al. (1991). The weakest observable shock effects in olivine are undulatory extinction and randomly oriented, nonplanar irregular fractures. Within increasing shock, olivine with undulatory extinction displays sets of parallel planar fractures with a spacing of tens of microns. Planar fractures are thought to be the most important shock indicators at moderate shock pressures. At some stage of shock intensity, olivine with planar fractures shows a distinctive mottled or mosaic appearance at extinction under the polarizing microscope. Single olivine crystals display numerous, more or less equant, poorly defined domains a few µm or less in size, which differ in their extinction positions by more than 3° to 5° of rotation. This causes the well-known shock-induced asterism observed in single crystal X-ray diffraction patterns. Olivine with strong mosaicism not only shows multiple sets of planar fractures but also additional planar deformation features (PDFs), which appear to be submicroscopic lamellae rather than open fractures. PDFs in olivine are thin, optical discontinuities which are about 20–40 µm in length. The spacing of PDF, mostly of the order of a few µm, is definitely narrower than that of planar fractures. Further increase of shock pressure leads to recrystallization of olivine in the solid state and, finally, melting. Recrystallized olivine is characterized by extremely fine-grained polycrystalline texture, which is accompanied by a yellow to yellow-brown staining of the recrystallized region. Typically, olivine grains adjacent to, or within, melt pockets and thick melt veins are often affected by recrystallization, which is restricted sometimes to the side of the grain located next to the melt vein. Recrystallized olivine lacks planar fractures and deformation features. It is assumed that these recrystallization textures are resulted from solid-state processes. At the stage of very strong shock, total melting of olivine takes place in regions adjacent to the recrystallized olivine. This olivine melt, however, coexists, simultaneously with shock-melted minerals, such as metal, troilite, plagioclase, and pyroxene. The "mixed" melt forms very fine-grained polycrystalline material with crystal sizes of the order of a few µm and less. In spatial connection with the formation of recrystallized olivine and of mixed polymineralic melt, olivine may transform into the crystalline high-pressure polymorph ringwoodite (Stöffler et al. 1991).

In the Suizhou meteorite, olivine is the most abundant constituent mineral. It holds about 45 % by volume. It is of white to light yellow or light green color in hand specimen and is colorless in thin section and is characterized by rather high refractive indices and strong birefringence. The occurrence of olivine in the Suizhou chondrite and its chemical composition and other properties are described in the following different sections.

Occurrence
Olivine in the Suizhou meteorite occurs as euhedral or semi-euhedral crystals. It has a grain size ranging from 0.05 to 0.5 mm. In some cases, the olivine grains are granulated with a grain size of less than 0.01 mm. This mineral displays wavy

extinction, irregular fractures, and from one to four sets of parallel planar fractures with a spacing of tens of microns (Figs. 3.1 and 3.2), but neither PDF nor solid-state recrystallization was observed in any of the Suizhou olivine grains (Xie et al. 2001a). However, some olivine grains show mosaic structure with subgrains of 150–200 μm in sizes, indicating that this mineral experienced a moderate shock metamorphic deformation (Fig. 3.3). Some elongated olivine grains with rounded tops and surrounded by plagioclase melt can also be observed in the Suizhou meteorite (Fig. 3.4). It was also revealed that fractures and cracks are more developed in the area adjacent to the both sides of the Suizhou shock veins that were formed during the cooling of shock veins, and some of the cracks are long enough to cut through the vein (Fig. 3.2).

SEM study in BSE mode on a Suizhou large rounded olivine grain shows that it contains very thin (~ 1 μm) exsolutions in the form of widely spaced lamellae

Fig. 3.2 BSE image occurrence of olivine (*Olv*) with abundant fractures and cracks in both sides areas of a shock vein (*Vein*) in the Suizhou meteorite. Note the fine-grained vein matrix inside the *rectangle*

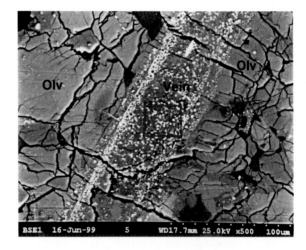

Fig. 3.3 Crossed polarized micrograph showing the mosaic structure in the Suizhou olivine (*Olv*). Note the subgrain domains in different optical extinction orientations and the clear domain boundaries. Plg = plagioclase (Chen et al. 2004a)

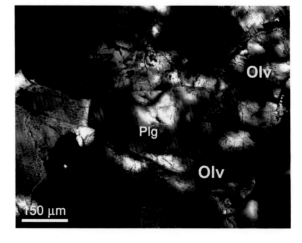

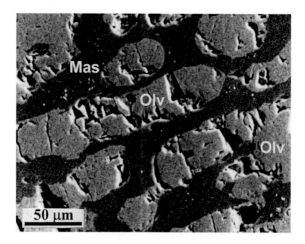

Fig. 3.4 BSE image showing the elongated olivine (*Olv*) grains with rounded tops surrounded by plagioclase melt (*Mas*). Note the tiny olivine fragments of 1–2 μm in size in plagioclase melt as mineral inclusions

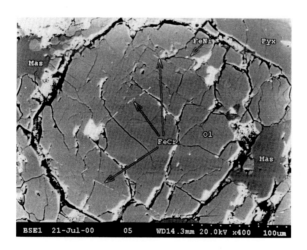

Fig. 3.5 BSE image showing the planar fractures and widely spaced exsolution lamellae of chromite (FeCr) oriented in three directions in a large olivine (*Ol*) grain; Pyx = pyroxene; Mas = maskelynite; FeNi = FeNi metal (Xie et al. 2011)

(Fig. 3.5). EDS analyses show that these lamellae contain FeO and Cr_2O_3, indicating that Fe and Cr were exsolved from olivine along (100) and (010) upon thermal metamorphism of the meteorite and crystallized in the form of exsolution chromite.

Chemical Composition

Olivine in type 6 ordinary chondrites is generally equilibrated in their major element compositions. Within each chondrite, the standard deviation on the mean Fa content is <1 mol and commonly <0.5 mol. Minor element contents of olivine in type 6 ordinary chondrites are very uniform. TiO_2, Al_2O_3, and Cr_2O_3 contents are all extremely low, <0.05 wt%. CaO contents are also very low, <0.1 wt%. MnO contents are between 0.4 and 0.5 wt% (Brearly and Jones 1998).

EPMA results of Suizhou olivine are shown in Table 3.1. From this table, we can see that the olivine has the chemical formula of $(Mg_{1.50}Fe_{0.49}Mn_{0.01})_2SiO_4$, and its chemical composition follows the above-mentioned rules for type 6 ordinary chondrites in the contents of all major and minor elements. The mean Fa content of Suizhou olivine is 24.12 mol, and the average contents of major elements SiO_2, MgO, and FeO of Suizhou olivine are 38.249, 39.174, and 22.343 wt%, respectively, and the standard deviations for Fa and major element contents are quite small, indicating that the olivine in the Suizhou chondrite is equilibrated in its major element composition.

The minor element contents of Suizhou olivine are very uniform. Al_2O_3 is absent in the composition of Suizhou olivine, and the TiO_2 and Cr_2O_3 contents are extremely low. All of them are <0.02 wt%. The CaO contents are also very low, only 0.011 wt%. The MnO contents are between 0.477 and 0.486 wt%. Na_2O and K_2O contents are also extremely low. They all are <0.02 wt%. It is interesting to

Table 3.1 Chemical composition of olivine in the Suizhou meteorite (wt%)

	SZ-Olv-1	SZ-Olv-2	SZ-Olv-3	Average	Olivine[a] average of 19	Olivine[b] average of 8
SiO_2	38.424	38.242	38.082	38.249	38.66	37.38
TiO_2	0.010	0.014	0.020	0.015	0.08	<0.01
Al_2O_3	0.000	0.000	0.000	0.000	0.51	<0.01
Cr_2O_3	0.003	0.020	0.017	0.013	0.51	n.d.
MgO	39.478	39.162	38.881	39.174	37.99	40.45
CaO	0.000	0.007	0.011	0.006	0.10	<0.01
MnO	0.486	0.480	0.477	0.481	0.52	0.15
FeO	22.296	22.324	22.410	22.343	21.76	22.35
NiO	0.079	0.038	0.000	0.039	0.12	n.d.
Na_2O	0.010	0.013	0.018	0.014	0.25	<0.01
K_2O	0.000	0.012	0.022	0.011	0.06	<0.01
P_2O_5	n.d.	n.d.	n.d.	n.d.	0.12	n.d.
Total	100.786	100.312	99.938	100.345	100.68	100.35
Fa	23.94	24.23	24.31	24.12	24.3	23.78
Fo	75.54	75.31	75.18	75.34	75.7	76.22
Mo	0.00	0.01	0.02	0.01		
Li	0.08	0.04	0.00	0.04		
Te	0.53	0.52	0.52	0.52		

All values were determined by EPMA in weight %; *n.d.* not detected; Fa—fayalite, Fo—forsterite, Mo—monticellite, Li—liebenbergite, Te—tephroite
[a]Data from Wang et al. (1990)
[b]Data from Hou et al. (1990)

point out that the Suizhou olivine contains small amount of NiO. Its average content is 0.039 wt%.

Raman Spectroscopy

Raman spectra recorded from the Suizhou olivine grains show strong and sharp peaks at 851 and 821 cm^{-1}, and weak peaks at 915, 951, 584, 424, and 299 cm^{-1} (Fig. 3.6) that are characteristic for a low-pressure phase of olivine. The both strong peaks at 851 and 821 cm^{-1} correspond to Raman modes of symmetric stretching vibration of the SiO$_4$ tetrahedra (v_1). The 915 and 951 cm^{-1} bands are attributed to the antisymmetric stretching vibration mode of the SiO$_4$ tetrahedra (v_3), and the 584 cm^{-1} peak corresponds to the antisymmetric deformation vibration mode of Si–O–Si (v_4), whereas the peak at 299 cm^{-1} can be attributed to the bending vibration mode of Mg–O (v_2).

X-ray Powder Diffraction Features

The X-ray powder diffraction patterns of the Suizhou olivine were obtained by using powder samples with three different sizes, namely 120–200 mesh, 200–350 mesh, and <350 mesh. The measurement conditions are as follows: Cu target, voltage 40 kV, current 30 mA, sampling width 0.020°, scan speed 20.00°/min, and scan axis 2θ/θ. The processing conditions are as follows: wave length 1.54056, and smoothing points 13. The obtained main diffraction lines are as follows: 2.4754 (10), 2.7895(9), 2.5308(9), 3.9208(5), 3.5228(5), 2.2862(5), 5.1751(4), 2.2630(4), and 2.1732(2). This indicates that this is a low-pressure phase of olivine, namely chrysolite.

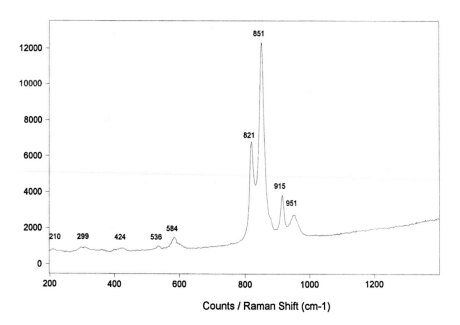

Fig. 3.6 Raman spectrum of olivine in the Suizhou meteorite

Table 3.2 Granularity of Suizhou olivine with different powder sizes (nm)

Powder size	Sample 1	Granularity (nm) Sample 2	Average
120–200 mesh	67.71	71.90	69.8
200–325 mesh	241.31	274.45	257.88
<325 mesh	66.90	69.94	68.42
Average	125.30	138.76	132.03

Another purpose of doing X-ray diffraction analysis is to detect the granularity of olivine grains after shock which can be calculated by using the measured half-amplitude width of the 2.789 Å peak. The more intense shock pressure the meteorite experienced, the smaller granularity of olivine will be. Table 3.2 shows the measured granularity of two Suizhou olivine samples with different powder sizes. From this table, we can see that the granularity of olivine in Suizhou meteorite ranges from 68 to 275 nm, and the average granularity is 132 nm. Comparison with the measured granularity of olivine in the heavily shocked Yanzhuang S6 chondrite (54 nm) and that of very weakly shocked Xinyang S1-S2 chondrite (>1000 nm), the olivine granularity of moderately to strongly shocked Suizhou S4-S5 chondrite is in between of them.

TEM Observations

Transmission electron microscopy (TEM) observations conducted on an ion-thinned Suizhou olivine sample by a Hitach transmission electron microscope revealed three types of shock-induced deformation features. One is the free edge dislocations, namely 3–4 sets of parallel edge dislocations, that can be clearly observed (Fig. 3.7). Second is domain boundary structures with the grain size of 2–3 μm in general (Fig. 3.7). It is interesting to point out that the edge dislocations developed only inside the domains and do not penetrate the domain boundaries. Third is the waving line reflecting the uneven optical extinction caused by shock compression.

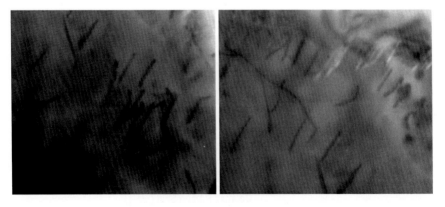

Fig. 3.7 Two bright field electron photomicrographs showing the edge dislocations and subgrain boundary structures in the Suizhou olivine crystals

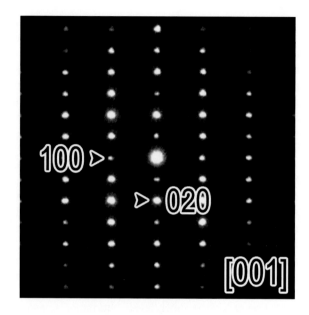

Fig. 3.8 Selected area electron diffraction pattern of the Suizhou olivine along [001] (after Zhang et al. 2006)

Figure 3.8 is the selected area electron diffraction (SAED) pattern of the Suizhou olivine showing rather regular diffraction spots in orthorhombic symmetry, with space group *Pbnm* and cell parameters of $a = 0.477$ nm, $b = 1.026$ nm, and $c = 0.60$ nm (Zhang et al. 2006). The absence of asterism for diffraction spots implies that olivine in the Suizhou meteorite experienced not so strong shock metamorphism.

3.2.2 Pyroxenes

Pyroxenes are significant phases in the Earth's upper mantle and are common silicate minerals in stone meteorites. Pyroxenes are metasilicates with chain structure which frequently form original rock constituents. The general formula for pyroxene can be expressed as XYZ_2O_6, where X represents Na, Ca, Mn^{2+}, Fe^{2+}, Mg, and Li in the distorted 6- to 8-coordinated M2 site; Y represents Mn^{2+}, Fe^{2+}, Mg, Fe^{3+}, Al, Cr, and Ti in the octahedral M1 site; and Z represents Si and Al in the tetrahedral site. Ferric iron also enters the tetrahedral site under certain bulk composition, temperature, pressure, and oxygen fugacity conditions. Chromium usually occurs as Cr^{3+} and titanium as Ti^{4+}, but under the reducing conditions that obtained in some meteorites, Cr^{2+} and Ti^{3+} may occur.

The members of low-Ca pyroxenes include enstatite ($Mg_2Si_2O_6$), ferrosilite ($Fe_2Si_2O_6$), orthopyroxene $(Mg,Fe^{2+})_2Si_2O_6$, and pigeonite $(Mg,Fe^{2+},Ca)_2Si_2O_6$. The low-Ca pyroxenes can be characterized chemically by En, Fs, and Wo in mol %, where En represents enstatite content, Fs represents ferrosillite content, and Wo represents wollastonite ($Ca_3Si_3O_9$) content.

Pyroxene structures are comprised of slabs of octahedral extending along the c-axis and connected top, bottom, left, and right to other octahedral slabs by single tetrahedral chains. As for the crystal structure of low-Ca pyroxenes, enstatite has three polymorphs with symmetries, $Pbca$, $P2_1/c$, and $Pbcn$, and ferrosillite has two polymorphs with symmetries, $Pbca$ and $P2_1/c$. Orthopyroxene only has one space group symmetry, $Pbca$, and pigeonite has two polymorphs with symmetries, $P2_1/c$ and $C2/c$. Hence, Magnesium iron pyroxenes that have $Pbca$ space group symmetry are referred to as orthopyroxene, whereas those that have $P2_1/c$ symmetry are referred to as pigeonite.

The shock effects in pyroxene in ordinary chondrites have been investigated by Stöffler et al. (1991). They found that in pyroxene, mostly orthopyroxene and diopside, all of the typical shock effects known from terrestrial, lunar, or experimentally shocked rocks have been observed in chondrites: Undulatory extinction, mechanical twinning, mosaicism and planar deformation features (PDFs), and melting are the major effects observed with increasing shock intensity. Mechanical polysynthetic twinning starts to form at pressures where olivine begins to develop planar fractures. PDFs in pyroxene coexist with strongly mosaicized olivine and maskelynite. In contrast to olivine, pyroxene is not affected by solid-state recrystallization at highest shock levels but, rather, transforms directly into molten state and gets mixed into the polyphase melt of veins and pockets (Stöffler et al. 1991). The transformation of pyroxene into the high-pressure polymorph majorite is also known.

Pyroxene is a main constituent mineral in the Suizhou meteorite, but it is not as abundant as mineral olivine. According to the EPMA and Raman microprobe analyses, the most common phase is low-calcium orthopyroxene, and only a few clinopyroxene grains were observed in the Suizhou chondrite.

Occurrence

Most of low-calcium pyroxene grains in the Suizhou meteorite have intact structures with a grain size of 0.05–0.8 mm. However, they usually contain abundant regular fractures and 3–4 sets of planar fractures (Figs. 3.1 and 3.9) and often display a weak mosaic texture and mechanical polysynthetic twinning. Optically, the undulatory extinction is very weak for most pyroxene grains. Shock-induced granulation (down to grain sizes of <0.02 mm) is also observed in some pyroxene grains, but no PDF or transformation into molten state was found in the Suizhou pyroxenes. Sometimes small pyroxene fragments can be observed in the melt pockets of plagioclase as mineral inclusions (Fig. 3.10).

Chemical Composition

Low-Ca pyroxenes in type 6 ordinary chondrites are generally equilibrated in their major element compositions, but they may have a wider range of Fs contents. In type 6 ordinary chondrites, essentially all low-Ca pyroxene is orthorhombic. Inversion of orthoenstatite to clinoenstatite may also be induced by shock, and some highly shocked type 6 ordinary chondrites contain twinned grains of low-Ca pyroxene derived from this process. Minor element contents of low-Ca pyroxene show significant range from chondrite to chondrite, and there is also significant

Fig. 3.9 BSE image showing
the fractured low-calcium
pyroxene grains in the
Suizhou unmelted chondritic
rock. Note the small rounded
grains of FeNi metal of taenite
composition (FeNi).
Mas = maskelynite

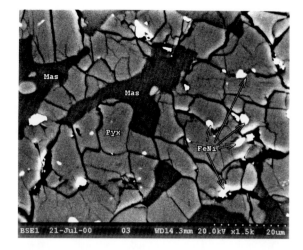

Fig. 3.10 BSE image
showing the regular fractures
in pyroxene (*Pyx*) and small
inclusions of pyroxene and
chromite (*Ch*) in a melt
pocket of maskelynite (*Mas*)
in the Suizhou meteorite (Xie
et al. 2006)

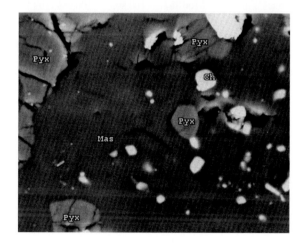

variation between type 6 chondrites, of 0.1–0.2 wt% in most minor elements. The
pigeonites have higher Cr_2O_3 than most low-Ca pyroxenes, 0.4–0.9 wt%. MnO
contents of low-Ca pyroxenes are similar to those in olivine, 0.4–0.5 wt% (Brearly
and Jones 1998).

EPMA results of Suizhou low-Ca pyroxene are shown in Table 3.3. From this
table, we can see that the low-Ca pyroxene has the chemical formula of
$(Mg_{0.79}Fe_{0.21}Ca_{0.02})_{1.02}(Si_{0.99}Al_{0.01})_{1.00}O_3$, and it follows the above-mentioned
rules for type 6 ordinary chondrites in their contents of major and minor elements.
The mean Fs content of Suizhou low-Ca pyroxene is 20.12 mol%; the average
contents of major elements SiO_2, MgO, and FeO of Suizhou low-Ca pyroxene are
55.775, 29.174, and 13.952 wt%, respectively; and the standard deviations for major
element contents are quite small, indicating that the low-Ca pyroxene in the Suizhou

Table 3.3 Chemical composition of low-calcium pyroxene in the Suizhou meteorite (wt%)

	SZ-Pyx-1	SZ-Pyx-2	SZ-Pyx-3	Average	Pyroxene[a] average of 10	Pyroxene[b] average of 21
SiO_2	55.926	55.761	55.639	55.775	55.76	55.6
TiO_2	0.199	0.129	0.157	0.162	0.17	
Al_2O_3	0.171	0.162	0.142	0.158	0.10	0.11
Cr_2O_3	0.092	0.128	0.110	0.110	0.20	0.19
MgO	29.193	29.283	29.435	29.239	28.18	29.6
CaO	0.714	0.753	0.678	0.715	0.65	0.83
MnO	0.461	0.529	0.497	0.496	0.52	0.44
FeO	13.765	13.947	14.143	13.952	13.59	14.3
NiO	0.000	0.003	0.019	0.007	0.07	
Na_2O	0.038	0.063	0.027	0.043	0.24	
K_2O	0.000	0.010	0.018	0.009	0.11	
P_2O_5	n.d.	n.d.	n.d.	n.d.	0.16	
Total	100.593	100.767	100.865	100.666	99.75	101.1
Fs	20.83	19.77	20.03	20.12	21.3	21.0
En	77.80	78.92	78.52	78.41	77.17	77.40
Wo	1.37	1.31	1.45	1.38	1.53	1.6

All values were determined by EPMA in weight %; *n.d.* not detected; Fs—ferrosite, En—enstatite, Wo—wollastonite
[a]Data from Wang et al. (1990)
[b]Data from Zeng (1990)

chondrite is also equilibrated in its major element composition. The minor element contents of Suizhou low-Ca pyroxene are very uniform. The TiO_2, Al_2O_3, and Cr_2O_3 contents are very low. All of them are <0.2 wt%. The CaO contents are as low as of 0.678–0.753 wt%. The MnO contents are between 0.461 and 0.529 wt%. The Na_2O contents are also very low. Its mean content is 0.043. The mean NiO content is extremely low, only 0.07 wt%. K_2O is almost absent in the Suizhou pyroxene.

Electron microprobe study also revealed 3 clinopyroxenes grains in the Suizhou chondrite. One is of pigeonite composition ($Wo_{6.5}En_{74.5}Fs_{19.0}$) inside a chondrule, and the other two are of diopside composition ($Wo_{45.3-46.5}En_{45.7-46.7}Fs_{6.8-9.0}$) (Zeng 1990).

Raman Spectroscopy

Raman analyses were conducted on the low-calcium pyroxene in Suizhou, and the spectra display sharp peaks at 335, 660, 679, and 1106 cm^{-1} (Fig. 3.11), indicating that this is a low-pressure phase of pyroxene. Our recent study has also observed two diopside grains in the Suizhou chondritic matrix. They have irregular shape and light gray color similar to that of orthopyroxene, but our micro-Raman spectroscopic study showed that these grains give Raman spectra (Figs. 3.12 and 3.13) identical to that of the standard diopside. Both grains show three strong peaks at 390–391, 664–666, and 1011–1012 cm^{-1} and a weak band at 322–324 cm^{-1}.

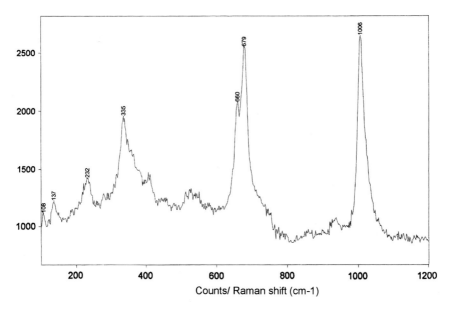

Fig. 3.11 Raman spectra of low-calcium pyroxene in the Suizhou meteorite

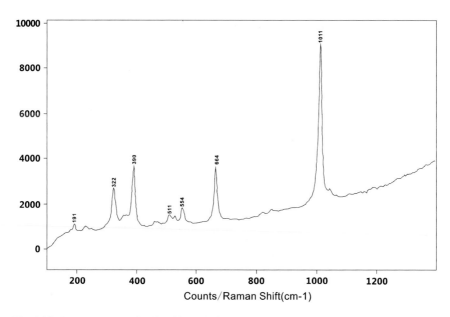

Fig. 3.12 Raman spectra of a diopside grain in the Suizhou meteorite

X-ray Powder Diffraction Features

The X-ray powder diffraction patterns of the Suizhou low-Ca pyroxene were obtained by using powder samples. The measurement conditions are as follows: Cu

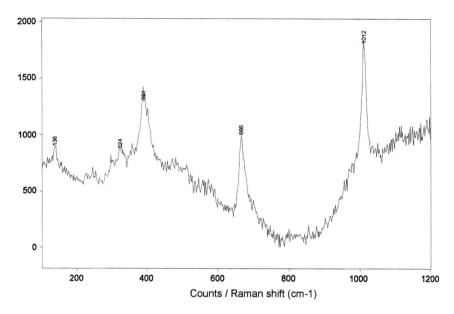

Fig. 3.13 Raman spectra of another diopside grain in the Suizhou meteorite

target, voltage 40 kV, current 30 mA, sampling width 0.020°, scan speed 20.00°/min, and scan axis $2\theta/\theta$. The processing conditions are as follows: wave length 1.54056 and smoothing points 13. The obtained main diffraction lines are as follows: 3.1907(10), 2.8878(9), 2.9588(4), 2.0971(3), and 2.8430(3). This indicates that this is a low-pressure phase of orthopyroxene.

TEM Observations

TEM observations conducted on an ion-thinned Suizhou pyroxene sample by a Hitach transmission electron microscope revealed that the main shock-induced deformation structure is the modulated structure in the form of parallel lamellae (Fig. 3.14) and the domain boundary structure with the grain size of 2–5 μm (Fig. 3.15). It is interesting to point out that the lamellae are generally penetrating the domain boundaries implying that the domain boundary structure appealed before the formation of lamellae.

Figure 3.16 is a high-resolution transmission electron microscopy (HRTEM) image of Suizhou pyroxene projected along [105] that shows anti-phase domain boundary (APB) marked by several crystallographic planes and fast Fourier transform (FFT) patterns shown in the inset (Zhang et al. 2006). Figure 3.17 displays two SAED patterns of the Suizhou orthopyroxene showing rather regular diffraction spots in orthorhombic symmetry, with space group *Pbca* and cell parameters of $a = 1.8278$ nm, $b = 0.8863$ nm, and $c = 0.5201$ nm (Zhang et al. 2006).

Fig. 3.14 Electron photomicrograph showing modulated structure in the form of parallel lamellae in the Suizhou pyroxene

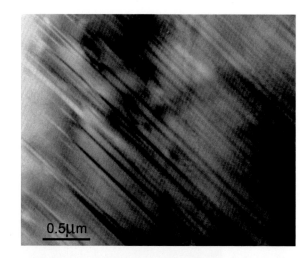

Fig. 3.15 Electron photomicrograph showing domain boundary structure with the subgrain size of 2–5 μm in the Suizhou pyroxene

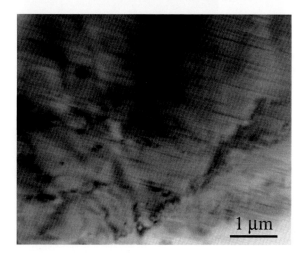

3.2.3 Plagioclase Feldspar

The feldspar group of minerals is the most important of the crustal phases because it makes up about 50–60 % of its total volume (Prewitt and Downs 1998). The feldspar group approximates a ternary system and may be considered in term of three components: albite, $Ab = NaAlSi_3O_8$; orthoclase, $Or = KAlSi_3O_8$; and anorthite, $An = CaAl_2Si_3O_8$. Or and Ab form the alkali feldspar group, with An absent or a minor constituent. Ab and An form the plagioclase group where a range in composition may occur from 100 % Ab to 100 % An. The minerals of feldspar group are widely distributed in ordinary chondrites, but they belong to the plagioclase group from albite to anorthite with minor content of celsian ($BaAl_2Si_3O_8$). Therefore, plagioclase is a common rock-forming mineral in L-group chondrites.

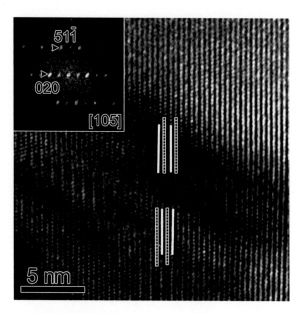

Fig. 3.16 High-resolution transmission electron microscopy (HRTEM) image of pyroxene projected along [105]. Anti-phase domain boundary marked by several crystallographic planes and fast Fourier transform patterns shown in the *inset* (after Zhang et al. 2006)

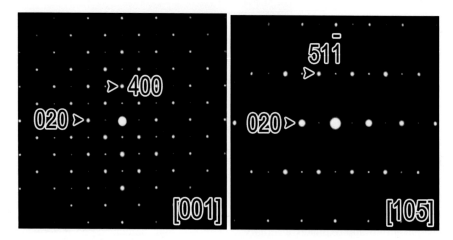

Fig. 3.17 Selected area electron diffraction patterns of the Suizhou orthopyroxene along [110] and [105] (after Zhang et al. 2006)

The crystal structures of all the feldspars show similar topology, and it can be considered as made up of four-membered tetrahedral rings that are linked together to form chains that run parallel to the *c*-axis. These chains form layers that are separated by channels. The layers are stacked with chains over chains to form pairs, and each pair is shifted to be over the channels of the adjoining pairs. The Na, K, and Ca atoms occupy the channels situated between layers within the pairs (Prewitt and Downs 1998).

According to the study of Stöffler et al. (1991), plagioclase in ordinary chondrites displays following shock effects with increasing shock intensity: plastic deformation as expressed by undulatory extinction at low shock; mosaicism, accompanied by sets of PDF and partial isotropization at moderate shock; complete transformation into diaplectic glass (maskelynite) at high shock; and melting and deformation of normal glass at highest shock. PDFs which have been shown to represent isotropic lamellae are sets of fine, μm-sized, planar lamellae that have an equal width and spacing of mostly less than 2 μm.

It is found that about 70–80 % of plagioclase grains in the Suizhou meteorite were melted and transformed into maskelynite, and only a small portion keeps the damaged plagioclase structure. So we will describe the mineralogical features of this remained plagioclase and those of maskelynite together in this chapter.

Occurrence

Plagioclase in Suizhou shows irregular shape and rather coarse-grained size. Plagioclase grains generally have normal optical properties indicating that its crystal structure remains intact (Fig. 3.18, upper image). However, most of the remaining plagioclase grains show a reduced birefringence, and some grains display a partially isotropic nature (Fig. 3.18, lower image). We also revealed that some plagioclase grains contain abundant PDFs in two to three sets (Fig. 3.19). A lot of plagioclase grains, especially those grains that are close to the shock veins, were transformed into maskelynite, a melted plagioclase glass.

Chemical Composition

Microprobe analyses (Table 3.4) show that plagioclase in the Suizhou meteorite has an oligoclase composition of $(Na_{0.77}K_{0.08}Ca_{0.11}Fe_{0.02})_{0.98} Al_{1.01}(Si_{2.91}Al_{0.09})_{3.00}O_8$. The chemical composition of maskelynite, $(Na_{0.73}K_{0.06}Ca_{0.10}Fe_{0.01})_{0.90} Al_{1.03}(Si_{2.92}Al_{0.08})_{3.00}O_8$, is close to that of plagioclase. The very low contents of Cr_2O_3, MgO, and FeO in maskelynite grains containing chromite and silicate fragments indicate that no exchange of elements has taken place between the host plagioclase melt and the trapped fragments of Cr-, Mg-, and Fe-bearing minerals within the plagioclase melt pockets.

Raman Spectroscopy

It has been revealed that most of the plagioclase grains in Suizhou meteorite have already transformed into maskelynite, and only a small portion of plagioclase grains is observed in thin sections. For comparing the structural change, the Raman spectra of maskelynite and plagioclase in the unmelted Suizhou chondritic rock, as well as that of unshocked plagioclase taken from Data Base, are shown in Fig. 3.20. It was revealed that the Raman spectra of plagioclase and maskelynite are quite different. The main peaks for unshocked plagioclase taken from Data Base are sharp enough at 508 and 1102 cm^{-1}. They correspond to Raman modes of Si–O–Si bending vibration and Si–O stretching of the SiO_4 tetrahedra, respectively. On the other

Fig. 3.18 Transmitted light
images of a chondritic portion
of the Suizhou meteorite. The
upper plane polarized image
shows the occurrence of
plagioclase (*Plg*). Note that
these plagioclases are cut by
many cracks. The *lower
image* is a crossed polarized
micrograph of the same area
as the *upper one*, showing
crystalline plagioclase.
Olv = olivine; M = FeNi
metal (Chen et al. 2004a)

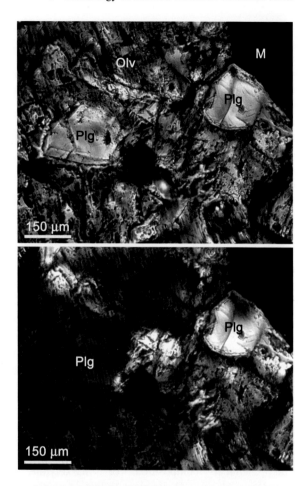

Fig. 3.19 A plane polarized
image showing abundant
planar deformation features in
2 sets in a plagioclase (*Plg*)
grain. Olv = olivine,
Pyx = pyroxene, Tr = troilite
(Chen et al. 2004a)

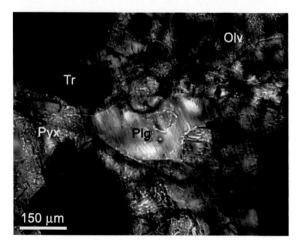

Table 3.4 Chemical composition of plagioclase in the Suizhou meteorite (wt%)

	SZ-Plg-1	SZ-Plg-2	SZ-Plg-3	Average	Plagioclase[a] average of 11	Plagioclase[b] average of 7
SiO_2	65.681	65.881	65.731	65.764	67.87	66.26
TiO_2	0.065	0.033	0.032	0.043	0.10	
Al_2O_3	21.373	21.934	21.993	21.767	21.19	21.46
Cr_2O_3	0.041	0.041	0.000	0.027	0.28	
MgO	0.000	0.013	0.000	0.004	0.33	0.04
CaO	2.169	2.162	2.295	2.209	2.20	2.16
MnO	0.009	0.000	0.041	0.017	0.07	
FeO	0.498	0.349	0.396	0.414	0.41	
NiO	n.d.	n.d.	n.d.	n.d.	0.11	
BaO	0.000	0.035	0.018	0.018	n.d.	
Na_2O	8.664	8.804	9.134	8.867	6.34	9.30
K_2O	1.380	1.311	1.248	1.313	1.09	1.12
P_2O_5	n.d.	n.d.	n.d.	n.d.	0.20	
Total	99.880	100.563	100.888	100.443	100.19	100.34
Ab	80.44	81.01	81.38	80.94		82.81
Or	8.43	7.94	7.29	8.89		6.55
An	11.13	10.99	11.30	11.14		10.64
Cs	0.00	0.07	0.03	0.03		

All values were determined by EPMA in weight %; *n.d.* not detected; Ab—Albite, Or—orthoclase, An—anorthite, Cs—Cesium
[a]Data from Zeng (1990)
[b]Data from Wang (1993)

hand, the Raman spectrum of maskelynite displays several broad bands in the ranges of 950–1250, 760–840, and 400–650 cm^{-1} (Fig. 3.20a), which are typical for glassy materials.

The Raman spectrum of Suizhou plagioclase shows sharp peaks at 508 and 475 cm^{-1} (Fig. 3.20b), implying that the crystal structure of plagioclase is still intact. However, instead of rather sharp peaks at 1102 and 765–817 cm^{-1} for unshocked plagioclase taken from Data Base (Fig. 3.20c), the Suizhou shocked plagioclase showed two broad bands at 1110 and 783 cm^{-1}, respectively. This implies that plagioclase in the Suizhou meteorite was heavily damaged by the shock event, although its crystal structure still keeps intact.

3.2.4 Maskelynite

Maskelynite is a form of amorphous plagioclase feldspar that was first described by Tschermak (1872). It was demonstrated to be a diaplectic glass formed by shock metamorphism when it was synthesized in a shock recovery experiment (Milton

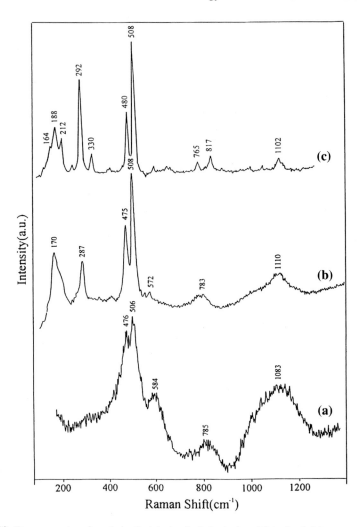

Fig. 3.20 Raman spectra of maskelynite (**a**), shocked plagioclase (**b**) in the Suizhou meteorite and that of normal plagioclase taken from the Data Base of Renishow Company (**c**)

and DeCarli 1963). Hence, the term of maskelynite was used in almost all known publications in the last 40 years as a synonym for diaplectic plagioclase glass. The term diaplectic glass was first introduced by Engelhardt et al. (1967), in analogy to experimentally shock-induced amorphous phases, to describe shock-induced solid-state transformations of quartz and feldspar to short-range order phases in shocked crystalline rocks in suevite of the Ries crater. In shocked meteorites, the formation of maskelynite is thought to be directly linked to the formation of PDFs in plagioclase, which represents partially transformed material, and in the melt regions of highly shocked samples, plagioclase has been described as normal glass, which was interpreted to have melted during shock. Stöffler et al. (1991) pointed out that maskelynite, the diaplectic oligoclase glass, forms clear transparent isotropic grains which perfectly retain their former crystal shape, whereas normal glass, as indicated by some flow-deformed shape, can be found only in very heavily shocked regions of chondrites as inclusions in melt veins and pockets, or adjacent to them.

It is interesting that under the shock-induced high pressures and temperatures, the silicate minerals in olivine and pyroxene in the Suizhou unmelted chondritic rock kept intact, but many of the plagioclase grains experienced melting and quenching and transformed into maskelynite.

Maskelynite, an important constituent of shocked meteorites, once thought to be diaplectic plagioclase glass is formed by shock-induced solid-state transformation. However, Chen and El Goresy (2000) reported that what appears to be maskelynite can actually be shock-melted plagioclase that quenched to glass at high pressures. Their systematic investigation of shocked L chondrites and ASNC meteorites indicates that maskelynite does not contain inherited fractures or cleavage, and shock-induced fractures, such as intragranular fractures. They found no evidence for models calling for melting that initiated in PDFs and affected whole crystals. Maskelynite grains are smooth and display radiating cracks emerging from their surface into neighboring pyroxene, which is indicative of shock-induced melting and quenching at high pressure, thus erasing inherited fractures and shock-induced fractures. This was followed by relaxation of the dense plagioclase glass, which induced the expansion cracks in pyroxene and olivine. They also indicate that enrichment in potassium and the lack of vesiculation in the melt pockets, melt veins, and molten mesostasis are clear evidence for melting and quenching under high pressure. They concluded that maskelynite in meteorite is not a diaplectic glass formed by solid-state transformation, but a quenched plagioclase melt.

Occurrence

Two occurrence types of maskelynite can be observed in the Suizhou unmelted chondritic rock. One is the thetomorphic maskelynite, an in situ melted glass with reserved grain morphology of the host plagioclase. This melt type has following specific features: isotropic under crossed polarized light, smooth grain surface, absence of inherited fractures or cleavage, shock-induced fractures, and lack of vesiculation and stoichiometric in chemical composition (Fig. 3.21). Some maskelynite grains are surrounded by radiating cracks, which penetrate deeply into the neighboring olivine and pyroxene as a result of volume expansion from the original crystalline phase to melt glass (Fig. 3.22). Another is the allomorphic maskelynite that fills fractures and cracks in other nearby silicate minerals, forming melt pockets or melt veinlets (Figs. 3.4, 3.8, 3.23 and 3.24). They were formed through injection of shock-induced plagioclase melt into fractures or cracks in the neighboring olivine and pyroxene. Sometimes, it occurs in the interstices around the olivine or pyroxene grains and interconnected with each other forming network of melt veinlets. These features imply that the plagioclase melt was formed under shock-produced high pressure and followed by migration of the melt for short distance under pressures. Furthermore, some larger plagioclase melt pockets

Fig. 3.21 Optical
micrographs showing a
completely isotropic
maskelynite (*Ms*) grain. The
upper image is a plane
polarized image; and the
lower one is a crossed
polarized image of the same
area. Note the surrounding
olivine displays mosaic
structure. Pyx = pyroxene
(Chen et al. 2004a)

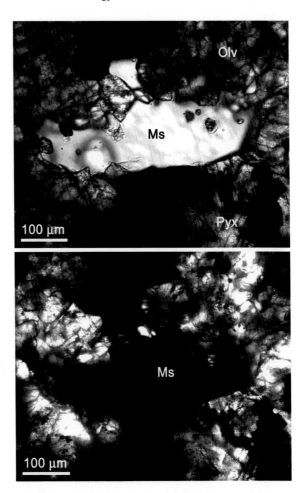

Fig. 3.22 BSE image
showing the radiating
fractures round the
thetomorphic maskelynite
(*Msk*) grains. Ol = Olivine
(Xie et al. 2001a)

Fig. 3.23 BSE image showing the occurrence of allomorphic maskelynite (*Msk*) on the both sides of a shock melt vein. Note the fragment inclusions of chromite in Msk. Ol = olivine, Pyx = pyroxene (Xie et al. 2011)

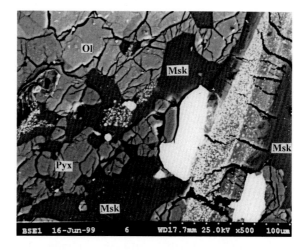

Fig. 3.24 BSE image showing the allomorphic maskelynite veinlets filling fractures in olivine (*Ol*) and pyroxene (*Pyx*), and maskelynite melt pockets containing tiny chromite fragments (*upper right corner*)

contain abundant small mineral fragments of chromite and, in less cases, pyroxene or olivine inclusions (Figs. 3.23 and 3.24). Under the optical microscope, this allomorphic-type maskelynite in Suizhou also displays a smooth surface and isotropic properties and contains no cleavages and fractures. It is clear that both two types of plagioclase glasses are similar in optical and physical properties but different in occurrences.

Chemical Composition

The results of microprobe analyses for both types of Suizhou maskelynite are shown in Table 3.5. It has been revealed that both type of plagioclase melts have the same chemical compositions as the plagioclase in Suizhou unmelted chondritic rock, and there is no obvious diffusion of chemical elements, such as Al, Cr, Mn, Na, and K, between the plagioclase melt and the surrounding silicate minerals or trapped in it mineral fragments.

Table 3.5 Chemical composition of maskelynite in the Suizhou unmelted chondritic rock (wt%)

	SZ-Msk-1	SZ-Msk-2	SZ-Msk-3	SZ-Msk-4	Maskelynite average	Plagioclase average
SiO_2	65.015	65.131	66.161	65.778	65.521	65.764
TiO_2	0.035	0.066	0.000	0.065	0.042	0.043
Al_2O_3	22.270	21.576	20.720	21.401	21.492	21.767
Cr_2O_3	0.000	0.011	0.000	0.022	0.008	0.027
MgO	0.145	0.886	0.506	0.965	0.625	0.004
CaO	2.139	2.454	2.211	2.059	2.216	2.209
MnO	0.000	0.047	0.047	0.000	0.025	0.017
FeO	1.091	1.176	0.925	1.127	1.080	0.414
BaO	0.000	0.000	0.000	0.000	0.000	0.018
Na_2O	9.300	8.401	9.302	9.533	9.134	8.867
K_2O	0.979	1.311	0.969	0.886	1.036	1.313
Total	100.974	101.059	100.841	101.836	101.179	100.443
An	83.59	81.82	83.34	84.71	83.37	80.94
Or	5.79	4.97	5.71	5.18	5.42	8.89
An	10.62	13.21	10.95	10.11	11.22	11.14
Cs	0.00	0.00	0.00	0.00	0.000	0.03

All values were determined by EPMA in weight %; *n.d.* not detected, Ab—albite, Or—orthoclase, An—anorthite, Cs—celsian

Raman Spectrometry

Raman spectroscopic investigations were conducted for both thetomorphic and allomorphic maskelynites in the unmelted Suizhou chondritic rock (Fig. 3.25). It has been revealed that the Raman bands for the Suizhou thetomorphic maskelynite are quite broad at 1083, 802, and 506 cm^{-1} (Fig. 3.25a), which are similar to the Raman spectra of the synthesized plagioclase glass (Fig. 3.25b). The Raman spectra of Suizhou allomorphic maskelynites (Fig. 3.25c, d) can be compared with that of thetomorphic maskelynite. The only difference in Raman spectra of allomorphic maskelynites is the presence of Raman peaks of olivine (850 and 819 cm^{-1}) or pyroxene (785, 1013, 667 and 390 cm^{-1}), for the allomorphic maskelynites contain a lot of inclusions of silicate minerals.

Plagioclase–Maskelynite Transformation Conditions

Shock wave experiments have shown that feldspar transforms into diaplectic glass at 26–34 GPa and into melt glass at 42 GPa (Ostertag 1983). However, the transition process may be accelerated at lower pressures when the temperature is high enough. For instance, the shock wave experiments of Schmitt (2000) indicate that plagioclase will transform into an amorphous phase at 20–25 GPa if the temperature in the target rock is elevated by about 920 K before the shock, whereas such a transformation at 293 K would take place at 25–30 GPa. It is known that an experimental shock pressure of 30 GPa at the ambient temperature of the chondritic rocks will only produce a temperature rise of less than 350 °C (Stöffler et al. 1991). Such a temperature is too low to melt plagioclase although the shock pressure is

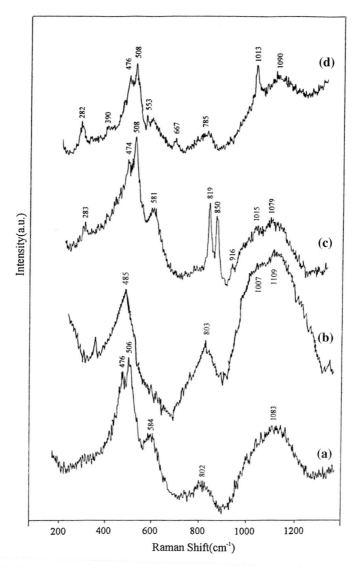

Fig. 3.25 Raman spectra of plagioclase glasses. **a** Thetomorphic maskelynite, **b** synthesized plagioclase glass, **c** allomorphic maskelynite with olivine inclusions, **d** allomorphic maskelynite with pyroxene inclusions (Xie et al. 2006)

very high. According to the normal noble gas contents (Wang 1993), we found no evidence that the Suizhou meteorite was heated to a higher temperature before the shock event. However, the maskelynite occurring in the Suizhou meteorite indicates that the shock-produced temperature must be far higher than 350 °C. The investigation on the basaltic meteorite Zagami by Langenhorst and Poirier (2000) revealed that the shock pressure and equilibrium shock temperature might reach

30 GPa and 1000 °C, respectively, since most of plagioclases in this meteorite were transformed into maskelynite (Langenhorst et al. 1991; McCoy et al. 1992; Chen and El Goresy 2000). Therefore, we assume that the plagioclase–maskelynite transformation conditions in the Suizhou meteorite might be in a pressure range of 25–30 GPa and at a temperature of about 1000 °C. According to the scheme of Stöffler et al. (1991), this pressure range corresponds to shock stage 4 (15–30 GPa).

3.3 Opaque Minerals

The major metallic phases in ordinary chondrites are FeNi metal, namely kamacite and taenite; tetrataenite is rare, and awaruite (with an empirically determined compositional range in meteorites of 63–72 wt% Ni) is very rare (Rubin 1997). The mean taenite/kamacite ratios in L-group chondrites are ~ 0.3. The major sulfide phase in ordinary chondrites is troilite; pentrandite occurs as an accessory phase in some oxidized LL-group chondrites. Chromite is the dominant oxide in ordinary chondrites; ilmenite and rutile occurs as rare phases. Metallic Cu is a rare but ubiquitous phase in ordinary chondrites; its modal abundance is 10^{-4} wt%, but it occurs in at least 66 % of the ordinary chondrites (Rubin 1997). In the Suizhou L6 chondrite, following metallic phases have been identified: FeNi metal, troilite, chromite, ilmenite, and metallic copper. Besides, the nanoscale pyrrhotite intergrowing with troilite is also observed in the Suizhou meteorite. Here, we introduce the physical and chemical properties of these six metallic minerals with the emphasis on their shock effects.

3.3.1 FeNi Metal

FeNi metal is one of the common opaque minerals in ordinary chondrites which contain 8–20 wt% of FeNi metal and occurs as grains typically 100–200 μm in size, evenly distributed throughout the meteorites. There are three mineral names for metallic iron–nickel phases in meteoritics, namely kamacite (α-FeNi), taenite (γ-FeNi), and plessite (fine-grained intergrowth of kamacite and taenite with Ni contents of 20–25 wt%). Kamacite in most equilibrated ordinary chondrites occurs as grains which are single crystals, but also be polycrystalline. Grains consisting of both kamacite and taenite are also present. In many chondrites, shock-produced twin lamellae, termed Neumann bands, are common. Kamacite grains have heterogeneous compositions with higher Ni content in their cores than at their edges. The actual contents vary depending on the size of the grain, but large grains are 6–7 wt% in the cores and 5–6 wt% within a few microns of the edges of the grains (Brearley and Jones 1998). Taenite commonly contains 25–35 wt% of Ni at the center of crystals and 45–55 wt% of Ni at the edges. Metallic phases in ordinary chondrites are particularly susceptible to the effects of shock metamorphism, both in the solid state and as a result of melting. Many chondrites have been reheated as a result of shock which has affected the metal in a number of different ways, including the formation of martensite (α_2-FeNi), disturbed M-shaped diffusion profiles, melting of metal (and troilite), the development of steep Ni gradients in martensite, the formation of secondary kamacite from taenite, chemical homogenization of metal grains, and enrichment of phosphorus in metal (Brearley and Jones 1998). In some shock-reheated chondrites, the FeNi metal contains elevated concentrations of P (>0.10 wt%) (Smith and Goldstein 1977), significantly higher than that in unshocked equilibrated chondrites (e.g. <0.01 wt%). In the Suizhou meteorite, FeNi metal is a major opaque mineral phase with abundance of 7–8 wt% and shows even distribution in this meteorite. Here, we describe the occurrence, chemical composition, and shock features of FeNi metal in the Suizhou meteorite.

Occurrence

FeNi metal in the Suizhou meteorite occurs as single grains of irregular shape and displays no obvious intragranular textures (Fig. 3.26). The grain size of FeNi metal ranges from 80 to 400 μm. Under reflect light, metal grains show white color and smooth surface, and some shock-induced fractures can be observed in them. However, neither shock-produced twin lamellae and formation of martensite (α_2-FeNi), nor melting of metal was observed in the Suizhou metal grains. This indicates that FeNi metal in the Suizhou meteorite had experienced only weak shock metamorphism. SEM observations in BSE mode have revealed some trace of shock-produced fragmentation in a FeNi metal grain of 200 μm in size which contains olivine and maskelynite as mineral inclusions (Fig. 3.27). It can still be seen that the left part of this metal grain was shock-granulated into grain size less than 5 μm, thus forming a unique pattern of micro-fractures.

It should be pointed out that some small rounded FeNi metal grains of 0.5–5 μm in diameter are also observed. However, they are unevenly distributed in the cracks or intersecting joints of shock-induced planar fractures in olivine and pyroxene (Figs. 3.9 and 3.28).

Fig. 3.26 BSE image showing FeNi metal (*FeNi*) and troilite (*FeS*) in the Suizhou chondrite. Olv = olivine; Pyx = pyroxene; Mas = maskelynite

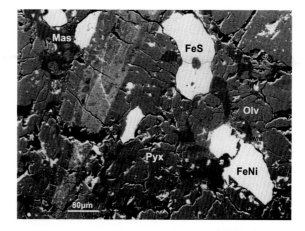

Fig. 3.27 BSE image
showing a large FeNi metal
grain containing olivine (*Olv*)
and maskelynite (*Mas*) as
inclusions. Note the trace of
fragmentation in *left area*

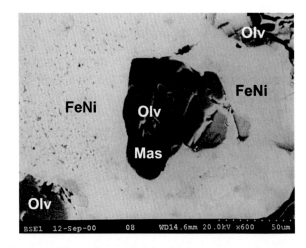

Fig. 3.28 BSE image
showing some small rounded
FeNi metal grains unevenly
distributed in the cracks or
intersecting joints of fractures
in olivine (*Olv*)

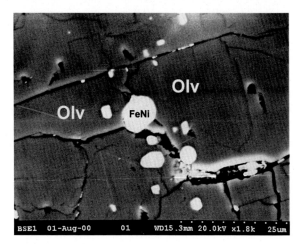

Chemical Composition

Electron microprobe analyses of 19 grains revealed that coarse-grained FeNi metal in the Suizhou chondritic rock has the composition of kamacite with the Fe content in the range of 91.15–96.54 wt%, the Ni content in the range of 4.40–6.81 wt%, the Co content in the range of 0.30–1.12 wt%, the mean Fe content in the range of 92.72 wt%, the mean Ni content in the range of 5.73 wt%, and the mean Co content in the range of 0.69 wt% (Shen and Zhuang 1990). Electron microprobe analyses of 24 FeNi metal grains in the Suizhou chondritic rock with Ni content higher than 7 wt% showed that they have the composition of taenite with the Fe content in the range of 42.16–89.75 wt%, the Ni content in the range of 8.21–56.06 wt%, the Co content in the range of 0.00–0.82 wt%, the mean Fe content in the range of 66.99 wt %, the mean Ni content in the range of 31.84 wt%, and the mean Co content in the range of 0.32 wt% (Shen and Zhuang 1990). The above data indicate that the chemical composition of kamacite in the Suizhou chondrite is rather homogeneous, but the chemical composition of taenite in this meteorite is heterogeneous. A few idiomorphic tetrataenite crystals with up to 30 μm in size were also found by SEM and EDS studies in the Suizhou meteorite, and the average Ni content is 52.5 wt% (Ge et al. 1990).

The chemical composition of small rounded metal grains in the cracks and intersecting joints of shock-induced planar fractures was determined using a LINK ISIS 300 X-ray energy dispersive spectrometer attached to SEM. The results show that this type of metal grains has a much lower Fe content (84.58–85.47 wt%) and higher Ni content (14.53–15.42 wt%) than the large metal grains in the chondritic rock. This indicates that the small rounded metal grains are taenite and they might have experienced chemical fractionation during shock-induced melting or evaporation. Therefore, we consider that this type of metal grain was deposited in cracks and fracture joints from nearby shock-induced Ni-rich metal melt or, more likely, from the vapor phase produced from Ni-rich metal during the shock event.

It is interesting to point out that terrestrial weathering of FeNi metal along some curved and spiral lines can be observed inside some metal grains forming curved rope-like bodies consisted of numerous black tiny grains of 5 μm in size. EDS analyses show that they are oxide of iron and nickel—(Fe,Ni)O. Figures 3.29 and 3.30 display the occurrence, and Table 3.6 shows chemical compositions of such oxide mineral. In Fig. 3.29, an Al-bearing kamacite grain of irregular shape (point 1) is in close contact with a taenite grain (point 2) which contains Ni up to 24.49 wt %. The EDS measurements at points 3, 4, and 5 show that their compositions are similar and have the average content of FeO = 94.16 % and NiO = 5.84 %. From the occurrence of (Fe,Ni)O, we argue that such curved or spiral lines might reflect the shock-induced intragranular micro-texture in FeNi metal grains.

Shock Deformation Features

According to the appearance of FeNi metal in the Suizhou meteorite and characteristics in its chemical composition, following shock deformation features can be described: (1) No obvious intragranular textures (shock-produced twin lamellae and

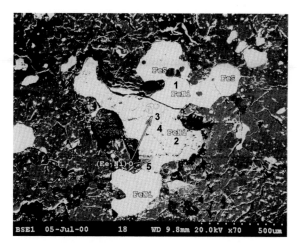

Fig. 3.29 BSE image showing terrestrial weathering of FeNi metal grain along some *curved lines*. Numbers *1–5* are EDS measurement points. FeNi = FeNi metal; FeS = troilite; (Fe,Ni) O = FeNi oxide

Fig. 3.30 Enlarged BSE
image of the central part of
Fig. 3.29 showing the *curved
rope*-like bodies consisted of
numerous *black tiny grains* of
(Fe,Ni)O in an FeNi metal
grain

Table 3.6 Compositions of (Fe,Ni)O in a Suizhou metal grain (wt%)

Oxides	Tainite (Point 1)	Kamacite (Point 2)	(Fe,Ni)O (Point 3)	(Fe,Ni)O (Point 4)	(Fe,Ni)O (Point 5)	(Fe,Ni)O average
FeO			94.64	94.17	93.67	94.16
NiO			5.31	5.83	6.35	5.84
Fe	75.51	90.64				
Ni	24.49	6.66				
Al		2.70				
Total	100.00	100.00	100.00	100.00	100.00	100.00

All data were measured by EDS technique

formation of martensite) were developed in the Suizhou metal grains; (2) rather
homogeneous chemical composition of kamacite in the chondrite; (3) no enrichment of phosphorus in metal grains; (4) neither partial melting at grain peripheries
nor melting of whole metal grains was observed in the chondrite; and (5) the
presence of a few small rounded Ni-rich metal grains observed in the cracks or
intersecting joints of shock-induced planar fractures in olivine and pyroxene is a
very rare case that was produced from the vapor phase during shock event.
Therefore, we assume that the FeNi metal in the Suizhou chondrite experienced
only weak shock metamorphism.

3.3.2 Troilite

Troilite is a ubiquitous constituent mineral in many meteorite types, lunar rocks,
and interplanetary dust particles (Kuebler et al. 1999; Skála and Císařová (2005);
Skála et al. 2006) and is one of the major opaque phases present in ordinary
chondrites. FeS group minerals are extraordinarily complicated due to their various

polymorphs and rich structural, magnetic, and electronic properties, which have attracted extensive studies. As is well known, all the FeS varieties with the general formula $Fe_{1-x}S$ ($0 \leqq x \leqq 0.125$) are formed based on the hexagonal NiAs-type substructure (with $A = 3.45$ Å and $C = 5.75$ Å), in which iron atoms occupy the octahedral interstices between the hexagonal closely packed layers of sulfur atoms. They include hexagonal troilite, monoclinic pyrrhotites, and hexagonal or intermediate pyrrhotites (Carpenter and Desborough 1964; Makovicky 2006). Detailed studies on several natural FeS minerals were performed by Morimoto et al. (1970), where the varieties with iron deficient Fe_7S_8, $Fe9S10$, $Fe_{10}S_{11}$, and $Fe_{11}S_{12}$ and stoichiometric FeS were termed 4C, 5C, 11C, 6C, and 2C (troilite) according to the number of unit C along the c-axis. Here, 4C corresponds to a monoclinic pyrrhotite variety and the others to hexagonal or intermediate pyrrhotites.

In equilibrated chondrites, troilite occurs as distinct grains, usually associated with FeNi metal and other opaque minerals. According to the summary of Brearley and Jones (1998), a large number of shock effects have been described in troilite in ordinary chondrites: (1) At relative low shock stage (S2), shock melting of sulfides can occur and submicron- to micron-sized troilite melt droplets can be present; and at higher shock stages (above S3), melting occurs along grain boundaries; and unmelted troilite grain edges, adjacent to the melt, develop a bubbly or swiss cheese texure. (2) Shock experiments show that below 10 GPa, troilite is monocrystalline, but becomes twinned between 10 and 20 GPa, and polycrystalline at shock pressures between 35 and 60 GPa. (3) Polycrystalline troilite is common in L chondrites which have been shocked to stage S4, and at stage S5, troilite becomes strongly sheared and polycrystalline. (4) TEM observations on troilite grains in Gaines County (H5) chondrite show that they contain high densities of dislocations, mosaicism, and planar lamellae. (5) The Ni content of troilite in shock-reheated chondrites was higher (<0.02–0.31 wt%) than in unreheated examples (<0.02 wt%), indicating that the Ni solubility in troilite decreases as a function of temperature. Troilite is one of main opaque minerals in the Suizhou meteorite with abundance of about 4.8 vol.% (Zeng 1990) and shows even distribution in this meteorite. Here, we describe the occurrence, chemical composition, Raman spectrum, transmission electron microscopic observations, and shock features of troilite mineral in the Suizhou meteorite.

Occurrence

Troilite in the Suizhou meteorite occurs mainly as single grains of irregular shape (Figs. 3.26, 3.31 and 3.32). The grain size of troilite ranges from a few micron to several hundred microns. Sometimes, troilite occurs in association with kamacite and taenite, or as veinlets filling the cracks in meteorite. Under reflect light, troilite shows light gray color and displays clear birefringence and strong anisotropic property. Fractures can be observed in many troilite grains; however, neither obvious intragranular textures, twinning, mosaicism, and planar lamellae, nor shock-induced melting was developed in the Suizhou troilites indicating that troilite in the Suizhou unmelted chondritic portion has only experienced rather weak shock metamorphism.

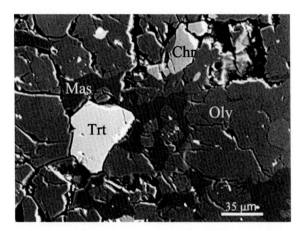

Fig. 3.31 BSE image showing the occurrence of troilite (*Trt*) and chromite (*Chr*) in the Suizhou meteorite. Note that surface of the troilite grain is rather smooth and no fractures are developed in the troilite grain. Olv = olivine, Mas = maskelynite

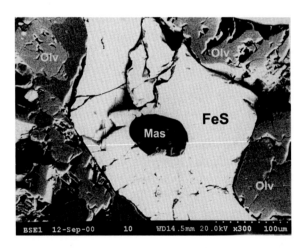

Fig. 3.32 BSE image showing the radial fractures and cracks in a large troilite (*FeS*) grain containing glassy plagioclase (*Mas*) as inclusion. Note the fractures are terminated at the boundaries of the host troilite grain with surrounding silicate minerals. Olv = olivine

SEM observations in BSE mode revealed that some large troilite grains containing glassy plagioclase as inclusions (Fig. 3.32) or surrounding glassy plagioclase bodies (Fig. 3.33) show some radial fractures or cracks around the glassy bodies. All the fractures or cracks are terminated at the boundaries of the host troilite grains with surrounding silicate minerals and with the glassy bodies. Such radial fractures or crack in troilite grains are generally thought to be formed by the volume increasing caused by the vitrification of plagioclase, while the troilite was

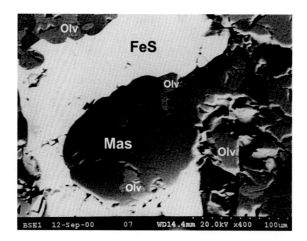

Fig. 3.33 BSE image showing the radial fractures and cracks in a large gulf-like troilite (*FeS*) grain surrounding a avoid glassy plagioclase (*Mas*) body. Note the radial fractures are also terminated at the boundaries of the troilite grain with surrounding silicate minerals. Olv = olivine

already quenched. This kind of fractures or cracks is the only intragranular texture observed so far in the troilite of the Suizhou meteorite.

Chemical Composition

Electron microprobe analyses of 16 grains revealed that coarse-grained troilite in the Suizhou unmelted chondritic rock has the composition of FeS with the Fe content in the range of 60.95–64.12 wt%, the Ni content in the range of 0.00–1.38 wt%, the Co content in the range of 0.00–0.48 wt%, the Cu content in the range of 0.00–0.58 wt%, the Ga content in the range of 0.00–0.67 wt%, the Ge content in the range of 0.00–0.34 wt%, and the S content in the range of 34.73–36.42 wt% (Shen and Zhuang 1990). The average contents are as follows: Fe—63.18 wt%, Ni—0.30 wt%, Co—0.09 wt%, Cu—0.239 wt%, Ga—0.14 wt%, Ge—0.16 wt%, and S—36.03 wt%, total 100.20 wt% (Shen and Zhuang 1990). It should be pointed out that the Ni content is less than 0.5 wt% for most of Suizhou troilite grains, and the higher Ni content was only observed for very few fine-grained troilites and for those tiny grains that occurred as inclusions in minerals.

Raman Spectroscopy

Raman spectrum of a troilite grain in Suizhou unmelted portion is shown in Fig. 3.34. From this figure, we can see rather sharp peaks at 283, 218, 398, and 600 cm^{-1}, a broad peak at 1292 cm^{-1}, and a weak peak at 456 cm^{-1}, indicating that troilite in the Suizhou meteorite has experienced rather weak shock deformation. This is consistent with the occurrence (lack of obvious intragranular texture) of this mineral in the Suizhou unmelted chondritic rock.

TEM Observations

Evans (1970) investigated the structure of troilite from lunar rocks, and the further structural aspects of transitions in stoichiometric FeS were described by Putnis

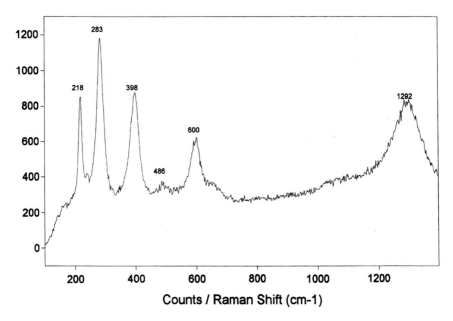

Fig. 3.34 Raman spectrum of troilite in the Suizhou meteorite

(1974), Besrest and Collin (1990a, b), Fleet (2006), and Ohfuji et al. (2007). 2C troilite, which is stable at room temperature and pressure, is a superstructure based on NiAs-type structure with space group $P\text{-}62c$, $a = 3\sqrt{A}$, $c = 2C$ (Fig. 3.35), and a cell content of 12 FeS. The unit cell shows small distortions from the ideal NiAs lattice positions. Cations in this structure are octahedrally coordinated, while anions are located at the centers of trigonal prisms. The octahedra share their faces parallel

Fig. 3.35 Schematic drawing of Fe positions of FeS group projected along c-axis showing the conventional unit cells of 2C troilite and 4C pyrrhotite superstructures derived from the NiAs-type structure (after Zhang et al. 2008)

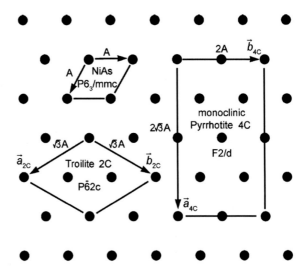

to the (001) plane and form infinite edge-sharing layers perpendicular to the hexagonal c-axis.

On the earth, metal-deficient pyrrhotite is more abundant than 2C troilite. Monoclinic 4C-type superstructure of pyrrhotite Fe_7S_8 with space group $F2/d$ and cell dimensions of $a = 3\sqrt{A}$; $b = 2A$, $c = 4C$, and $\beta = 90.43°$ (Fig. 3.35), occurring in nature, was originally determined by Bertaut (1953) and further refined by Tokonami et al. (1972). The 4C monoclinic pyrrhotite can be derived from 2C troilite by subtracting one eighth of the iron atoms. The resulting structure can be expressed as (...FAFBFCFDF...), which contains alternating layers free of vacancies (denoted as F) and vacancy-bearing layers (denoted as ABCD depending on the vacancy position). The superstructure further relaxes with lowering of the symmetry from the ideal hexagonal structure to slight distorted monoclinic one, in which the c-axis tilts a little with respect to the ab-plane (Wang and Salveson 2005).

Figure 3.36 displays the SAED patterns of the Suizhou troilite showing the regular round diffraction spots in hexagonal symmetry, with space group P-62c and cell parameters of $a = 0.5962$ nm and $c = 1.175$ nm (Zhang et al. 2006). It should be pointed out that some of diffraction spots in the Fig. 3.36 appeal slightly elongated implying a weak shock-induced lattice deformation of the Suizhou troilite.

Zhang et al. (2008) also investigated in detail the microstructures of FeS varieties existing in the Suizhou meteorite using TEM techniques with high spatial resolution including SAED combined with corresponding kinematic simulations and HRTEM. They reported that the primary hexagonal 2C troilite phase is found to intergrow with the minor monoclinic 4C pyrrhotite phase as nanometer scale domains, and furthermore, the APB with the displacement vector $1/4[001]_{2C}$ in 2C troilite and 60° rotation twinning in 4C pyrrhotite is revealed, which can be used to follow the lowering of the symmetry during the formation of ordered superstructures when cooling. The results in the following section are further discussed by Zhang et al. (2008) in the context of group theory (Töpel-Schadt and Müller 1982), with

Fig. 3.36 The selected area electron diffraction pattern of the Suizhou troilite along [0001] (after Zhang et al. 2006)

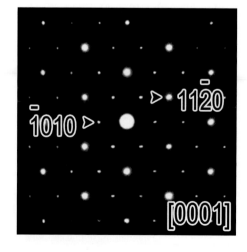

consideration of the shock and thermal metamorphism history experienced by the
Suizhou meteorite.

(i) **Intergrowth of 2C hexagonal troilite and 4C monoclinic pyrrhotite**

The intergrowth of 2C hexagonal troilite and 4C monoclinic pyrrhotite exists
throughout the FeS in the Suizhou meteorite as revealed by systematic SAED
examination. A typical SAED pattern (Fig. 3.37a) along the zone axis $[120]_{2C}$(//
$[110]_{4C}$) with its corresponding schematic illustration (Fig. 3.37b) clearly shows the
coexistence of both 2C troilite and 4C pyrrhotite phases. This is further confirmed
by the good agreement between the experimental pattern (Fig. 3.37a) and the
composite simulated SAED pattern (the inset in Fig. 3.37a) superimposed for both
phases. After image contrast enhancement, the weak diffraction spots characteristic

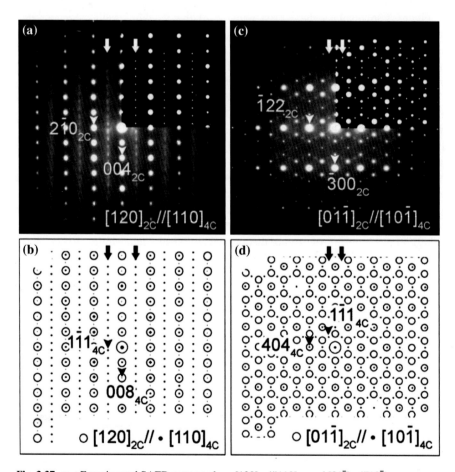

Fig. 3.37 a, c Experimental SAED patterns along $[120]_{2C}$//$[110]_{4C}$ and $[01\bar{1}]_{2C}$//$[10\bar{1}]_{4C}$ zone axes,
respectively, where the insets are the simulated SAED patterns with kinematic theory of electron
diffraction. **b, d** Schematic representations corresponding to the SAED patterns (**a, c**), respectively,
where *open circles* represent 2C troilite and *solid dots* 4C pyrrhotite (after Zhang et al. 2008)

of 4C pyrrhotite, two columns of which are indicated by the arrows, show obviously its presence besides the primary 2C troilite.

HRTEM was used to confirm the intergrowth of 2C hexagonal troilite and 4C monoclinic pyrrhotite down to the nanoscale. Figure 3.38a shows the HRTEM image taken along the same axis $[120]_{2C}$. The framed parts were enhanced using Fourier filtering with periodic masks for 4C and 2C, respectively, where the planes (001) of 4C pyrrhotite with d-spacing 2.27 nm and (001) of 2C troilite with d-spacing 1.16 nm are marked. This confirms the fine-scale intergrowth of both phases as nanometer domains. Similar to the SAED pattern in Fig. 3.37a, the corresponding FFT pattern (Fig. 3.38b) shows the coexistence of both phases as well. The intergrowth agrees with the former literature (Carpenter and Desborough 1964; Morimoto et al. 1975; Putnis 1975; Dádony and Pósfai 1990; Weber et al. 2006). Our structural analysis shows much weaker diffractions from 4C pyrrhotite than

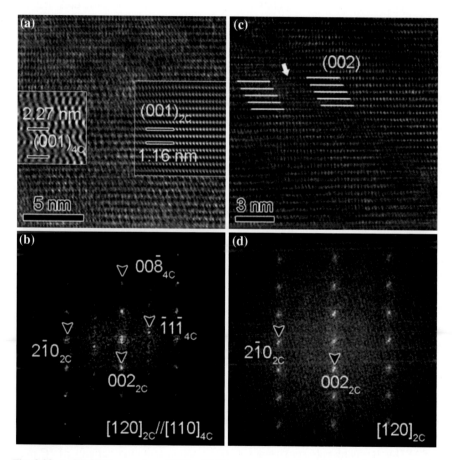

Fig. 3.38 a HRTEM image along $[120]_{2C}//[110]_{4C}$ zone axes, where the framed parts are Fourier filtered with periodic masks for 4C pyrrhotite and 2C troilite, respectively. **b** FFT pattern for image (**a**). **c** HRTEM image of 2C troilite along $[120]_{2C}$. **d** FFT pattern for image (**c**) (after Zhang et al. 2008)

those from 2C troilite even after selected image contrast enhancement, indicating that the content of 4C pyrrhotite is much less than 2C troilite in the meteorite. This agrees with previous EPMA and XRD observations (Wang and Li 1990).

As demonstrated by the above SAED and HRTEM analyses, FeS contains the fine-scale intergrowths of both monoclinic 4C pyrrhotite and hexagonal 2C troilite superstructures with a fixed orientation relationship. This may form through ordering from high-temperature NiAs-type structure with higher symmetry during cooling in the Suizhou meteorite. As pointed out, the formation of monoclinic 4C pyrrhotite needs somewhat slower cooling rates because rapid cooling of Fe_7S_8 would lead to the formation of a well-ordered hexagonal 3C structure (Groves and Ford 1963; Ericsson et al. 1997; Wang and Salveson 2005). In rocks that have undergone long cooling 4C pyrrhotite superstructure occurs with troilite (Weber et al. 2006). Weber et al. (2006) proposed an interpretation of the relationship between superstructures and temperatures. At high temperatures, the vacancy distribution is random and the unit cell of pyrrhotite is the NiAs subcell (Fig. 3.35). With decreasing temperature, an ordering process takes place to reduce the stress in the crystal until the minimum internal energy is reached when the vacancies have maximum separation: The crystal adopts a more ordered superstructure (Fig. 3.35). Furthermore, Wang and Li (1990) proposed that 2C troilite and 4C pyrrhotite were magnetized at 300–400 °C during the slow cooling process of the Suizhou meteorite.

(ii) **Anti-phase domain boundary in 2C troilite**

Further HRTEM inspection, also taken along $[120]_{2C}$ zone axis as shown in Fig. 3.38c, reveals a distinct APB in 2C troilite as labeled by a white arrow between the two nanoscale domains with crystal planes (002) indicated. Corresponding FFT pattern (Fig. 3.38d) clearly shows the direction of $[120]_{2C}$ for 2C troilite. Apparently, the APB (Fig. 3.38c) has a displacement vector of $1/4[001]_{2C}$.

In general, the subgrain domains as by-product will form with ordering from symmetry breaking indicated by group theory. 2C troilite superstructure ($P\text{-}62c$) might display APBs with displacement vectors $1 = 1/3\langle\bar{1}10\rangle_{NiAs}$ or $1/2[001]_{NiAs}$, which were theoretically deduced from the loss of translational symmetries from the NiAs-type structure ($P63/mmc$) (Töpel-Schadt and Müller 1982). The APB with the displacement vector $1 = 1/3\langle\bar{1}10\rangle_{NiAs}$ has been observed there (Töpel-Schadt and Müller 1982). The displacement vector $1/4[001]_{2C}$ of the present APB is just another one, i.e., $1/2[001]_{NiAs}$, which complements the experimental determination as required by group theoretical deduction.

(iii) **Twinning in 4C monoclinic pyrrhotite**

Following the group theory consideration, 60° rotation twinning domains about the pseudohexagonal c-axis will be present in the more ordered 4C pyrrhotite super-structure ($F2/d$) due to the further rotation symmetry reduction. Further examination of SAED patterns reveals coexistence of 60° rotation twinning variants of 4C pyrrhotite and the primary 2C troilite. A typical experimental SAED pattern along $[24\bar{1}]_{2C}//[33\bar{1}]_{4C}//[3\text{-}3\bar{1}]_{4C}$ zone axes (Fig. 3.39a) clearly shows diffraction spots

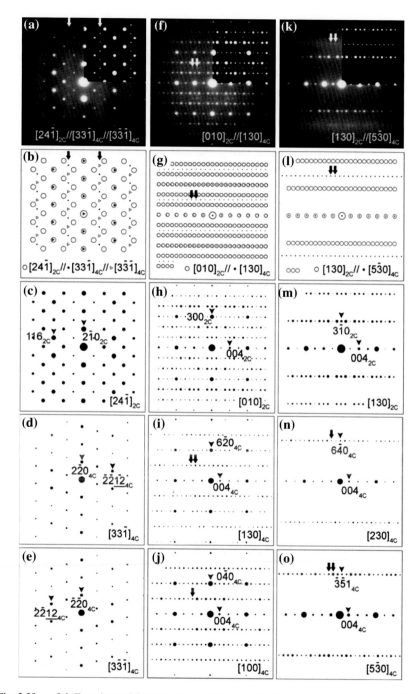

Fig. 3.39 **a, f, k** Experimental SAED patterns with *insets* the simulated composite patterns. **b, g, l** Schematic illustration corresponding to the SAED patterns (**a, f, k**). **c, h, m** Simulated patterns for 2C troilite, respectively. **d, i, n** Simulated patterns for 4C pyrrhotite, respectively. **e, j, o** Simulated patterns for 4C pyrrhotite 60° twinning variant, respectively (after Zhang et al. 2008)

arising from three origins, i.e., the 2C troilite ($[24\bar{1}]_{2C}$) variant, one 4C pyrrhotite variant ($[33\bar{1}]_{4C}$), and another 4C pyrrhotite 60° rotation twinning variant ($[33\bar{1}]_{4C}$), which is schematically illustrated in Fig. 3.39b. Kinematic simulated SAED patterns for the three variants separately are presented in Fig. 3.39c–e in sequence. Meanwhile, the inverted composite simulated SAED pattern with the three variants coexisting (inset in Fig. 3.39a) agrees with the experiment very well, which confirms the existence of the rotation twinning. The observation of twinning confirms the reports from the literature of defects characteristic of 4C monoclinic pyrrhotite, which includes twinning, stacking faults, and other unidentified defects (Morimoto et al. 1970; Bennett et al. 1972; Tokonami et al. 1972; Nakazawa et al. 1975; Putnis 1975; Kontny et al. 2000; Pósfai et al. 2000).

Besides that the primary 2C troilite variant always presents in SAED patterns, the occurrence of 4C or 4C 60° twinning pyrrhotite variants is observed to vary at the nanometer scale depending on the location within the specimen, which indicates that the 4C and 4C 60° twinning pyrrhotite variants are nanoscale domains. For example, one typical experimental SAED pattern along $[010]_{2C}//[130]_{4C}$ (Fig. 3.39f) shows the 4C pyrrhotite matrix variant (Fig. 3.39i) coexisting with the primary 2C troilite variant (Fig. 3.39h) without the 4C pyrrhotite 60° twinning variant (Fig. 3.39j). This can be revealed through careful inspection of the experimental SAED pattern (Fig. 3.39f) with consideration to the delicate differences between the simulated pattern for 4C pyrrhotite matrix variant (Fig. 3.39i) and that for 4C pyrrhotite 60° twinning variant (Fig. 3.39j), which are indicated by arrows. Another example (Fig. 3.39k–o) shows the 4C y and ordering pyrrhotite 60° twinning variant coexisting with the primary 2C troilite variant without the 4C pyrrhotite matrix variant. It is also noticed that the diffraction spots completely or to some extent coincide with each other between the 4C pyrrhotite 60° twinning variant and the 4C matrix variant. For example, in Fig. 3.37a, the diffraction spots from 4C pyrrhotite matrix variant $[110]_{4C}$ coincide with those from 4C 60° twinning $[1\bar{1}0]_{4C}$ completely, which cannot be differentiated. For another example, in Fig. 3.37c, the number of diffraction spots from 4C pyrrhotite matrix variant $[13\text{-}2]_{4C}$ is less than that from 4C 60° twinning $[10\bar{1}]_{4C}$ for different systematic kinematic absence, which raises uncertainty about the existence of the 4C matrix variant. The rotation twinning in the 4C superstructure may result from the ordered arrangements of metal vacancies. Ordering can take place by solid-state transformation from precursor structures. This viewpoint agrees with Kontny et al. (2000). This involved phase transition may be associated with the thermal metamorphism of the meteorite.

In summary, detailed TEM examination reveals the fine-scale intergrowth of the minor 4C monoclinic pyrrhotite and the primary 2C troilite superstructures of FeS in the Suizhou meteorite. APBs are present in the 2C troilite phase with the displacement vector $1/4[001]_{2C}$, which originate from the loss of translational symmetries from the high-temperature NiAs-type structure as expected by group theory analysis. Further rotational symmetry breaking causes the 60° twinning domains in more ordered 4C monoclinic pyrrhotite. The characteristic nanoscale intergrown microstructure was formed during the cooling of the Suizhou meteorite with spontaneous broken symmetry.

3.3.3 Chromite

Chromite is a common opaque mineral with spinel structure in ordinary chondrites. Based on textures and assemblages, Ramdohr (1967, 1973) recognized 6 types of chromites in ordinary chondrites: (1) coarse chromite; (2) clusters of chromite aggregates; (3) pseudomorphous chromite; (4) exsolution chromite; (5) chromite chondrules, and (6) myrmekitic (or symplectic) chromite. El Goresy (1976) pointed out that the coarse chromite type occurs in the overwhelming majority of ordinary chondrites, achondrites, stony irons, and iron meteorites as coarse euhedral to subhedral grains interlocked in the silicate matrix, and the clusters of chromite aggregates type seems to be restricted to ordinary chondrites. The cluster consists of medium- to fine-grained idiomorphic to subhedral grains of chromite usually embedded in plagioclase. The chromite chondrules in ordinary chondrites are rare, and they are enriched in chromium and contain >13 wt% of Cr_2O_3 in their bulk compositions. The pseudomorphic chromite type appears to be restricted to chondrules in ordinary chondrites, and the myrmekitic chromite type seems to be restricted to mesosiderites and iron meteorites. On the basis of a survey of 76 equilibrated H, L, and LL chondrites, Rubin (2003) claimed that in shocked equilibrated ordinary chondrites, some chromite occurs in chromite–plagioclase assemblages that were referred to as "clusters of chromite aggregates" by Ramdohl. It has been reported more recently that the abundance of chromite in ordinary chondrites increases with increasing petrologic type, and the mean composition also varies with petrologic type and chromite compositions become more homogeneous as metamorphic grade increases (Brearley and Jones 1998). The Suizhou meteorite is an L-group chondrite. Its petrologic type is 6, which is indicative of that the chromite abundance in this meteorite should be high enough, and the chromite composition should be more homogeneous.

Occurrence
Chromite is a common accessory mineral in the Suizhou meteorite, and this meteorite contains ∼1.4 % of chromite by volume which is indicative of that the chromite abundance in this type 6 chondrite is high enough. In the Suizhou unmelted chondritic rock, chromite occurs either in the form of single coarse grains or as tiny fragments embedded in molten plagioclase. The exolution chromite is extremely rare. We only found one case of lamellar chromite in an olivine crystal (Xie et al. 2011).

The first occurrence of chromite in the Suizhou unmelted chondritic rock is in the form of individual coarse euhedral, subhedral to irregular grains with grain size of about 50 μm, and, in some cases, euhedral chromite crystals up to 100 μm can be observed (Figs. 3.40 and 3.41). Sometimes, several chromite grains are gathered to form an aggregate (Fig. 3.42). The coarse chromite grains in the Suizhou meteorite display more abundant irregular and subparallel fractures than the neighboring pyroxene and olivine, and 2–4 sets of planar fractures can be recognized in them (Figs. 3.40 and 3.43). Figure 3.44 is a BSE image showing a chromite grain adjacent to a shock vein contains abundant cracks and fractures, in which 3–4 sets of planar fractures can be recognized.

Fig. 3.40 BSE image
showing a fractured chromite
grain in the Suizhou
meteorite. Note the irregular
cracks in this grain.
Olv = olivine;
Pyx = pyroxene;
Mas = maskelynite (Xie et al.
2001a)

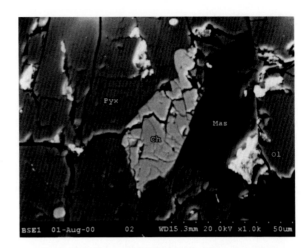

Fig. 3.41 BSE image of an
euhedral chromine grain in
the Suizhou meteorite. Note
the subparallel fractures and a
few irregular cracks in this
grain (Chen et al. 2003; Xie
et al. 2011)

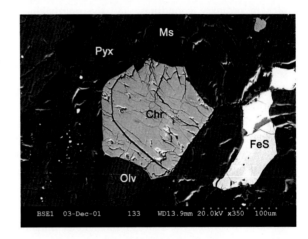

Fig. 3.42 BSE image
showing chromite (*Chr*)
grains associated with olivine
(*Olv*), pyroxene (*Pyx*), FeNi
metal, and troilite (*FeS*). Note
the chromite grains are
heavily fractured (Xie et al.
2011)

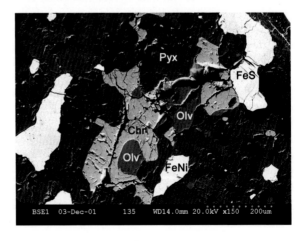

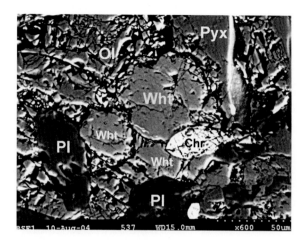

Fig. 3.43 BSE image showing an ovoid chromite (*Chr*) grain surrounded by whitlockite (*Wht*) grains. Note the loosely spaced planar fractures in this chromite grain. Pyx = pyroxene; Plg = plagioclase (Xie et al. 2011)

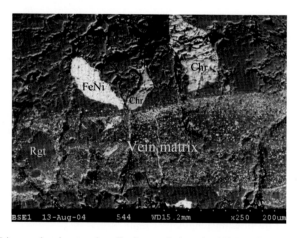

Fig. 3.44 BSE image showing two heavily fractured chromite (*Chr*) grains of irregular shape in the area adjacent to a shock vein in Suizhou meteorite. FeNi = FeNi metal; Rgt = ringwoodite

Figure 3.45 shows two heavily shocked and fractured coarse chromite grains with several sets of shock-induced planar fractures, and the upper right one has been shock-disaggregated into many small fragments with grain size ranging from 0.2 to 30 μm in length, and some of the larger chromite fragments still keep the original orientation, but many of the smaller fragments have become disoriented. Furthermore, a glide line crossing this chromite grain and some migration of chromite fragments up to 30 μm along this glide line can be observed. It should be pointed out that no zoning occurs in any of the chromite grains that we studied under optical microscope and SEM.

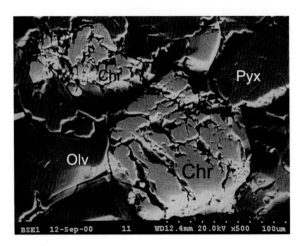

Fig. 3.45 BSE image showing two heavily fractured chromite (*Chr*) grains with several sets of shock-induced planar fractures in the Suizhou meteorite. Note the upper chromite grain has been shock-disaggregated into many small fragments. Olv = olivine, Pyx = pyroxene (Xie et al. 2011)

The second occurrence of chromite in the Suizhou unmelted chondritic rock is in the form of clusters of tiny fragments embedded in molten plagioclase (maskelynite). Here, we describe three different types of chromite fragments embedded in three plagioclase melt pockets, namely pocket 1[#] (Fig. 3.46), pocket 2[#] (Fig. 3.47), and pocket 3[#] (Fig. 3.48), in which the occurrences of tiny chromite fragments are rather different.

Figure 3.46 demonstrates the plagioclase melt pocket 1[#] in which a cluster of chromite fragments is trapped but not homogeneously distributed in the molten plagioclase. The chromite fragments range from 0.4 to 10 μm in sizes. Most of the tiny fragments in this type melt pocket have rounded shape and smooth surfaces, and some grain outlines and intrafractures of the host chromite grain are still preserved in some larger fragments. This implies that these tiny chromite fragments very likely came from a shock-disaggregated host chromite grain and were subsequently trapped by flowing plagioclase melt.

The second type of melt pockets is shown in Fig. 3.47 as pocket [#]2, where chromite grains embedded in molten plagioclase were smashed into grain size less than 0.5 μm forming a unique pattern of cluster consisted of hundreds of tiny fragments in the molten plagioclase (Fig. 3.47a). We observed a large variation of the grain size ranging from 0.2 to 5 μm for these chromite fragments (Fig. 3.47b). Although this type of chromite occurrence is uncommon, it can still be observed in the Suizhou meteorite. Another similar case is shown in the l bottom-left corner of Fig. 3.48a, where the melt pocket is in the elongated form, but the occurrence of embedded cluster of chromite is similar to that in melt pocket [#]2. We assume that such specific occurrence of embedded chromite in molten plagioclase might be referred to as "cluster of chromite aggregates" of Ramdohr, that is a deformed chromite–plagioclase chondrule (Ramdohr 1973).

Fig. 3.46 BSE images showing molten plagioclase (*Mas*) melt pocket [#]1 (*black*) containing clusters of chromite fragments (*white*). **a** Tiny chromite fragments in the molten plagioclase pocket. **b** Enlarged image of the central part of (**a**). Note the original intrafractures in 3 larger chromite (*Ch*) fragments. Ol = olivine, Pyx = pyroxene, FeNi = metal, Vein = shock melt vein (Xie et al. 2011)

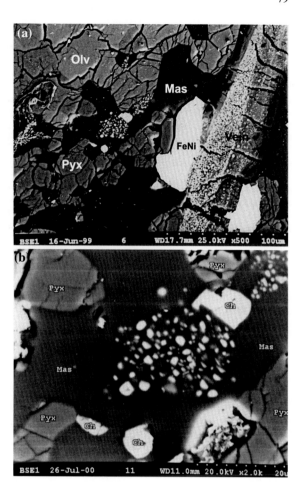

The third type of melt pockets is shown in Fig. 3.48 as pocket 3[#], in which some of elongated chromite fragments of 3–5 μm in length and 1–2 μm in width show preferential orientation (Fig. 3.48a, b). It should be pointed out that this is an only one case that we observed in the Suizhou meteorite. This phenomenon might be considered at first sight as showing that chromite slabs recrystallized from the plagioclase melt. However, the much higher melting point of chromite (∼2000 °C) compared with oligoclase and the easy fragmentation of chromite during shock imply that the present orientation of fragments was not produced by melting and recrystallization. Rather, it is due to shock-induced breaking and disaggregation of the host chromite grains along well-developed subparallel intrafractures in two sets and subsequent rapid in situ mixing with intruding molten plagioclase. It is interesting that some larger chromite fragments of quadrate and rhombic form and 5–9 μm in sizes are also observed in association with the small elongated chromite fragments (Fig. 3.48b).

Fig. 3.47 BSE image
showing the occurrence of the
highly granulated chromite
fragments (*white*) in the
molten plagioclase (*Mas*)
pocket [#]2. **a** The full view of
pocket [#]2. **b** Enlarged image
of the central part of image
(**a**). Note the rounded,
quadrate, and rhombic grain
outlines of the tiny chromite
fragments, and the large
variation of the grain sizes
ranging from 0.2 to 5 μm.
Ol = olivine, Pyx = pyroxene,
FeNi = metal (Xie et al. 2011)

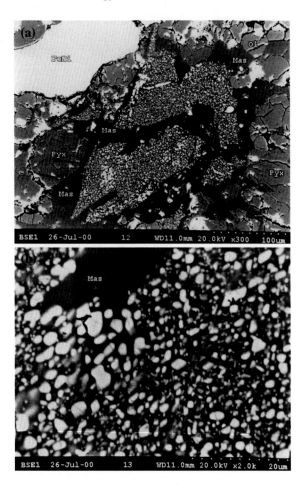

The exsolution chromite in ordinary chondrites is rare. This type is generally encountered in clinopyroxene-rich chondrules as fine-grained clusters of chromite at the boundaries between clinopyroxene and plagioclase. Ramdohr (1967) proposed exsolution from a chromium-rich silicate to account for this assemblage. However, we did not find this type of chromite in the Suizhou meteorite, but we observed one case in which an euhedral olivine crystal of 250 μm in size contains chromite lamellae in the form of fine straight intermittent threads of 1–1.5 μm in width and 50–100 μm in length, and very loosely spaced and oriented mainly along {100} of olivine (Fig. 3.5). This would be the third occurrence of chromite in the Suizhou unmelted main body. The host olivine crystal is weakly shock-fractured. EDS analysis shows that this host olivine has the composition of SiO_2 40.55, MgO 35.88, and FeO 23.57 wt%, which is close to the average composition of olivine in the Suizhou meteorite (Xie et al. 2001a).

Fig. 3.48 BSE images showing the molten plagioclase (*Mas*) melt pockets and veinlets in the Suizhou meteorite. **a** Highly granulated chromite fragments (*white*) in the melt pockets. Note the melt pocket [#]3 at the *upper right* of the image that contains elongated chromite fragments preferentially oriented in two directions. **b** Enlarged image of the melt pocket [#]3 that contains preferentially oriented chromite inclusions. Note some larger chromite fragments of quadrate and rhombic forms in the pocket. Ol = olivine; Pyx = pyroxene; Ch = chromite (Xie et al. 2011)

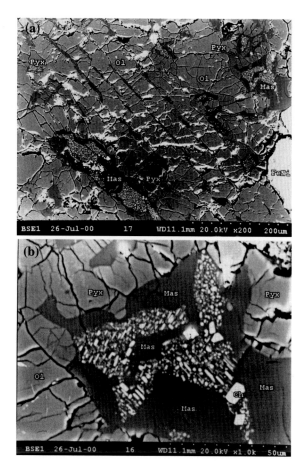

Chemical Composition

Snetsinger et al. (1967) reported that their results of electron microprobe analysis of chromite in seven "equilibrated" chondrites showed that within a given meteorite, chromite grains are remarkably similar in composition, and no zoning occurs. From meteorite to meteorite, however, the mineral has the narrow compositional range (wt%): Cr_2O_3 54.6–57.0, Al_2O_3 5.2–6.3, TiO_2 2.08–2.98, FeO 30.5–33.9, MgO 1.54–2.48, MnO 0.65–1.05, and V_2O_3 0.65–0.77, indicating that these chondritic chromites are higher in iron and titanium and lower in aluminum than most terrestrial chromites (Snetsinger et al. 1967).

The results of electron microprobe analyses of 8 single chromite grains in the Suizhou unmelted main body are shown in Table 3.7 (Xie et al. 2011). Among them, the first 5 grains were measured by us and the last 3 were given by Wang and Li (1990). From this table, we can see: (i) All 8 single chromite grains are remarkably similar in their chemical composition; (ii) from grain to grain, the mineral has a rather narrow compositional range (wt%), namely Cr_2O_3 56.52–58.12, Al_2O_3 5.15–6.38, TiO_2 2.30–3.11, FeO 29.7–31.71, MgO 2.09–2.73, MnO

Table 3.7 Compositions of single coarse chromite grains in the Suizhou meteorite (wt%)

Grain No	1	2	3	4	5	6[a]	7[a]	8[a]	Ave
MgO	2.52	2.60	2.73	2.58	2.67	2.09	2.46	2.33	2.50
FeO	29.77	29.63	29.27	29.60	29.35	31.71	30.39	30.04	29.97
MnO	0.73	0.83	0.91	0.81	0.99	0.48	0.88	0.59	0.78
CaO	n.d.	n.d.	n.d.	n.d.	n.d.	0.12	0.10	0.00	0.03
TiO_2	2.96	2.81	3.11	3.06	2.83	2.30	2.76	2.66	2.81
Cr_2O_3	57.01	56.62	57.06	56.64	56.74	56.60	56.52	58.12	56.93
Al_2O_3	6.17	6.26	6.38	6.37	6.26	6.06	5.15	5.69	6.10
V_2O_3	0.87	0.93	0.90	0.95	0.89	1.02	1.03	0.95	0.94
Total	100.03	99.96	99.68	100.36	99.73	100.38	99.29	100.38	100.06

All data were measured by EPMA; *n.d.* not detected
[a]From Wang and Li (1990)

0.48–0.99, and V_2O_3 0.87–1.03; (iii) the average composition of the Suizhou single chromite grains is falling in the compositional range given by Snetsinger et al. (1967) for chromites in "equilibrated" chondrites on the whole, and the only difference for the Suizhou chromite is in the slightly lower content of FeO (29.97 vs. 30.5–33.9 wt%) and higher content of V_2O_3 (0.94 vs. 0.65–0.77 wt%).

The EDS analysis has been used to investigate the possible variation in chemical composition of the tiny chromite fragments of different sizes embedded in different plagioclase melt pockets of the Suizhou meteorite. For comparison, the EDS analysis was also conducted for plagioclase in the Suizhou unmelted main body and for maskelynite in melt pockets (Xie et al. 2011).

It is revealed by EDS analyses that the chemical compositions of different size chromite fragments trapped in the three plagioclase melt pockets, e.g., the pocket [#]1 showed in Fig. 3.46b, the pocket [#]2 showed in Fig. 3.47b, and pocket [#]3 showed in Fig. 3.48b, are very close to each other (Tables 3.8, 3.9, and 3.10). All of them are very similar to those of coarse-grained chromite in the Suizhou unmelted main body showed in Table 3.7. We also analyzed the chemical compositions of chromite fragments embedded in other four plagioclase melt pockets located in different

Table 3.8 Compositions of chromite fragments embedded in the plagioclase melt pocket [#]1 (wt%)

Grain size (μm)	11 × 9	7 × 6	3.5 × 3	3 × 3	Average
MgO	2.6	2.6	3.1	2.9	2.8
FeO	30.4	30.4	30.3	30.1	30.3
TiO_2	2.8	3.1	3.3	2.9	3.0
Al_2O_3	5.7	5.3	5.2	6.1	5.5
Cr_2O_3	57.6	57.8	57.3	57.3	57.5
V_2O_3	0.9	0.8	0.8	0.7	0.8
Total	100.0	100.0	100.0	100.0	

All data were measured by EDS technique

Table 3.9 Compositions of chromite fragments embedded in the plagioclase melt pocket $^\#$2 (wt%)

Grain size (μm)	5 × 4	5 × 4	5 × 3	4 × 4	3.5 × 3	3 × 3	Average
MgO	3.1	2.5	3.1	2.5	3.0	2.9	2.9
FeO	30.2	30.3	30.0	30.2	30.1	30.6	30.2
TiO$_2$	2.9	2.4	2.9	2.3	2.9	2.4	2.6
Al$_2$O$_3$	5.7	6.1	6.0	5.5	6.0	6.1	5.9
Cr$_2$O$_3$	57.4	57.7	57.1	57.8	57.3	57.2	57.4
V$_2$O$_3$	0.7	1.0	0.9	0.7	0.7	0.8	0.8
Total	100.0	100.0	100.0	100.0	100.0	100.0	

All data were measured by EDS technique

Table 3.10 Compositions of chromite fragments embedded in the plagioclase melt pocket $^\#$3 (wt%)

Grain size (μm)	9 × 6	6 × 6	3 × 3	3 × 3	Average
MgO	3.2	3.0	3.0	3.1	3.1
FeO	30.1	30.0	30.1	30.3	30.1
TiO$_2$	2.9	3.1	2.8	2.9	2.9
Al$_2$O$_3$	5.3	5.9	6.1	5.5	5.8
Cr$_2$O$_3$	57.5	57.2	57.2	57.3	57.3
V$_2$O$_3$	1.0	0.8	0.8	0.9	0.9
Total	100.0	100.0	100.0	100.0	

All data were measured by EDS technique

distances from shock veins, e.g. pockets 4$^\#$, 5$^\#$, 6$^\#$, and 7$^\#$, by EDS technique, and found that their compositions are also very close to each other (Table 3.11) and similar to that of individual coarse chromite in the Suizhou unmelted main body. Our EDS results also showed that the chemical compositions of maskelynite (plagioclase melt) in these three melt pockets are almost identical, and all of them are similar to that of plagioclase in the Suizhou unmelted main body (Table 3.12).

Table 3.11 Compositions of chromite grains in other plagioclase melt pockets (wt%)

Oxides	Melt pocket 4$^\#$ 4 mm from vein	Melt pocket 5$^\#$ 2 mm from vein	Melt pocket 6$^\#$ 1 mm from vein	Melt pocket 7$^\#$ 1 mm from vein	Average
MgO	3.0	3.1	3.3	2.7	3.0
FeO	29.8	29.6	27.9	30.3	29.4
TiO$_2$	2.8	2.3	1.6	3.2	2.5
Al$_2$O$_3$	6.2	6.6	7.9	5.0	6.7
Cr$_2$O$_3$	57.3	57.7	58.6	58.0	57.9
V$_2$O$_5$	0.9	0.7	0.7	0.8	0.8
Total	100.0	100.0	100.0	100.0	

All data were measured by EDS technique

Table 3.12 Compositions of maskelynites in the molten plagioclase pockets (wt%)

Oxides No	Pocket [#]1 3	Pocket [#]2 3	Pocket [#]3 3	Maskelynite average	Plagioclase 3
SiO_2	67.7	68.8	68.4	68.3	68.1
CaO	2.4	2.4	2.4	2.4	2.4
Al_2O_3	20.8	20.3	20.6	20.6	20.6
Na_2O	8.3	7.4	8.0	7.9	7.8
K_2O	0.8	1.1	0.6	0.8	1.1
Total	100.0	100.0	100.0	100.0	100.0

All data were measured by EDS technique. *No* Number of analysis

The above fact implies that both maskelynite and chromi have the same compositions of plagioclase and chromite in the Suizhou unmelted main body, respectively, and no exchange in chemical elements between these two phases appeared during shock melting. The similar contents of Cr_2O_3, FeO, SiO_2, and Al_2O_3 in both type of chromites (individual coarse grains and fragments in plagioclase melt) support our assumption on the shock-disaggregated origin of the randomly and preferentially oriented chromite fragments mixed with plagioclase melt shown in Figs. 3.46b and 3.48b, respectively.

The chemical compositions of exsolution chromite in a Suizhou olivine crystal measured by EDS on four points are shown in Table 3.13. Since the chromite threads are thinner than 1.5 μm, the EDS results reflect a mixture composition of chromite and host olivine. It is interesting that analyses at point 1 and point 2 gave very similar results with no Al_2O_3 present in them, and those at point 3 and point 4 also show rather close results but with remarkable content of Al_2O_3. Hence, we assume that the composition of exsolution chromite in olivine in the Suizhou chondrite is variable in Al_2O_3 and not so homogeneous in composition. Our EDS analysis shows that the host olivine has the composition of SiO_2 40.6, MgO 35.9, and FeO 23.5 wt%, which is close to the average composition of olivine in the Suizhou unmelted main body (Xie et al. 2001a). Since the average Cr_2O_3 content in the Suizhou olivine is very low (only 0.03 wt%) and no Al_2O_3 was detected in olivine (Xie et al. 2001a), it is reasonable to assume that the chromite exsolution in olivine is an extremely rare case, and it might be formed for long period of time by thermal metamorphism, and not to be related with the shock metamorphism of the meteorite.

Table 3.13 Composition of exsolution chromite in an olivine crystal (wt%)

	Point 1	Point 2	Point 3	Point 4	Olivine
SiO_2	40.6	25.7	25.8	18.6	21.6
MgO	35.9	25.5	26.4	16.4	15.8
FeO	23.5	26.5	26.5	28.3	27.2
Cr_2O_3	n.d.	21.3	21.3	32.7	31.5
Al_2O_3	n.d.	n.d.	n.d.	4.0	3.9
Total	100.0	100.0	100.0	100.0	100.0

All data were measured by EDS technique. *n.d.* not detected

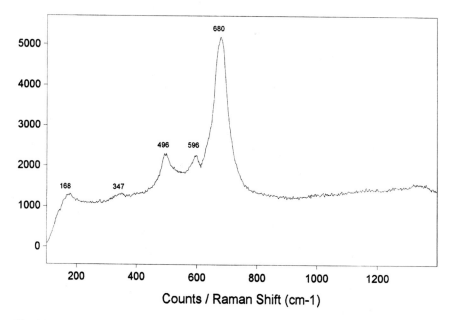

Fig. 3.49 Raman spectrum of chromite in the Suizhou meteorite

Raman Spectrometry

Raman spectrum of chromite in the Suizhou unmelted rock is shown in Fig. 3.49. From this figure, we can see an intense Raman peak at 650 cm^{-1}, two rather weak peaks at 486 and 218 cm^{-1}, and two weak peaks at 347 and 165 cm^{-1}. It should be pointed out that all the Raman peaks are rather broad. This indicates that chromite in the Suizhou meteorite has also experienced some shock-induced deformation. This is consistent with the development of subparallel and irregular fractures and cracks in this opaque mineral.

Fig. 3.50 Selected area electron diffraction pattern of the Suizhou chromite (after Zhang et al. 2006)

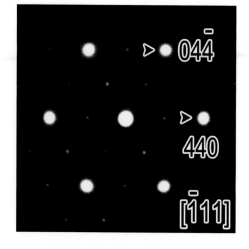

TEM Observations

TEM observations on the Suizhou chromite were conducted by Zhang et al. (2006). Figure 3.50 is the SAED pattern of the Suizhou chromite $FeCr_2O_4$ showing the round diffraction spots in cubic symmetry, with space group $Fd3m$ and the cell parameter 0.8379 nm. The absence of asterism for diffraction spots implies that chromite in the Suizhou meteorite experienced not so strong shock metamorphism.

3.3.4   Ilmenite

In type 6 ordinary chondrites, ilmenite is one of the main accessory opaque minerals. It has been revealed that ilmenite occurs in association with chromite, metal, sulfide, and rarely with aluminous spinel. It has the composition range 51–54 wt% of TiO_2, 35–45 wt% of FeO, 1–7 wt% of MgO, 0.6–1.0 wt% of MnO, and 0.03–1.4 wt% of Cr_2O_3. The Suizhou meteorite is one of the type 6 ordinary chondrites, and some ilmenite grains have been observed in this meteorite as one of its accessory opaque minerals. However, the ilmenite content is only about 0.1 % by volume which is about ten times lower than that of chromite. Furthermore, the grain size of ilmenite is also much smaller than that of chromite.

Occurrence

Ilmenite in the Suizhou meteorite occurs in association with troilite and/or other opaque minerals as small grains of irregular shape with grain size less than 20–30 μm. We could only observe two ilmenite grains in this meteorite. Figure 3.51 shows a small ilmenite grain of 20 μm × 30 μm in size which occurs in close contact with a large troilite grain. Its surface is rather smooth, and only one short

Fig. 3.51 BSE image showing the occurrence of a small ilmenite (*Imt*) grain in the Suizhou unmelted chondritic rock. FeS = troilite; Olv = olivine; Pyx = pyroxene; Mas = maskelynite; Apt = chlorapatite

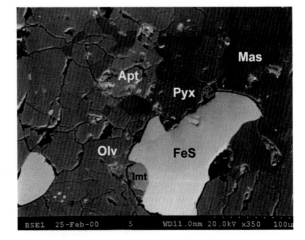

fracture is seen in this grain. Another ilmenite grain of 10 μm × 15 μm in size was observed inside a troilite grain.

Chemical Composition
The chemical composition of the two ilmenite grains that we observed in the Suizhou meteorite was analyzed by EDS technique. The results for the first ilmenite grain shown in Fig. 3.51 are as follows (in wt%): FeO—40.01 %, MgO—3.33 %, MnO—2.33 %, and TiO_2—54.33 %. Hence, its calculated chemical formula is $(Fe_{0.82}Mg_{0.12}Mn_{0.05})_{0.99}Ti_{1.00}O_3$. The EDS analysis for the second ilmenite grain that we observed as a inclusion in troilite has given the following similar results (in wt%): FeO—41.87 %, MgO—3.10 %, MnO—2.04 %, and TiO_2—52.98 %. The calculated chemical formula of this ilmenite is $(Fe_{0.87}Mg_{0.11}Mn_{0.04})_{1.01}Ti_{1.00}O_3$. The above results are in good consistence with that of EPMA on the Suizhou ilmenite sample S2 conducted by Shen and Zhuang (1990). They gave the average composition for two ilmenite grains as follows (in wt%): FeO—40.28, MgO—3.71, MnO—1.81, TiO_2—53.34, CaO—0.07, ZnO—0.04, SiO_2—0.09, Al_2O_3—0.08, V_2O_5—0.23, and Cr_2O_3—0.93. Its calculated formula is $(Fe_{0.82}Mg_{0.14}Mn_{0.04})_{1.00}(Ti_{0.98}Cr_{0.01})_{0.99}O_3$. This implies that ilmenite in the Suizhou meteorite is rather homogeneous in its chemical composition.

Shock Deformation Features
Under reflected light, the Suizhou ilmenite grains are gray in color and show rather smooth surface. There are only a few fractures developed in these grains, indicating that ilmenite in the Suizhou chondritic rock experienced very weak shock-produced deformation.

3.3.5 Natural Copper

Natural copper is one of the opaque accessory minerals in ordinary chondrites. Rubin (1994) documented 9 different occurrences of natural copper in ordinary chondrites. The two most common are as follows: (1) at kamacite–troilite interface and (2) adjacent to small troilite grains inside Ni-rich metal. In almost all cases, the natural copper grains are irregular in shape and have grain sizes between ∼1 and 6 μm, although rare, unusually large grains have also been described (Olsen 1973). Limited electron microprobe data indicate that the natural copper contains ∼1.5–2.0 wt% of Ni in solid solution (Olsen 1973; Rubin 1994). Unfortunately, no grains of natural copper were observed in the Suizhou meteorite by our microscopic and SEM studies. However, a few natural copper grains were found by Chen (1990) and Shen and Zhuang (1990) in this meteorite. Their descriptions about the occurrence and chemical composition of this mineral are as follows.

Occurrence
Natural copper occurs in the Suizhou meteorite either as small mineral inclusion in kamacite or taenite, or as individual grains in the interstices of kamacite, taenite and

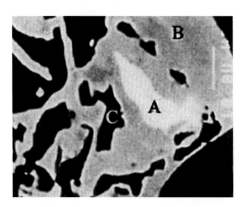

Fig. 3.52 BSE image showing three-phase system of natural copper (*A*, *white*), taenite (*B*, *light gray*), and troilite (*C*, *black*). The *vertical scale bar* on the *upper right corner* is 10 μm long (after Chen 1990)

troilite and, in rare cases, in the interstices of kamacite, taenite, and silicate minerals. Figure 3.52 is a BSE image of a leaflike natural copper grain embedded in taenite with the grain size of 8 μm × 27 μm. Under reflect light, natural copper shows violet-red color, isotropic optical property, and granular or irregular shape.

Chemical Composition

Natural copper of terrestrial origin usually contains up to 2–3 % of Au, 3–4 % of Ag, and 2–3 % of Fe. In meteorites, natural copper contains mainly Fe and Ni and does not contain Au and Ag. Electron microprobe analysis for the natural copper in the Suizhou meteorite conducted by Chen (1990) gives following results in wt%: Cu—93.22, Fe—4.53, Ni—2.25. Shen and Zhuang (1990) also conducted electron microprobe analysis for another copper grain and got following results in wt%: Cu —91.63, Fe—5.14, Ni—2.29, Ga—0.02, Ge—0.37, and S—0.18. Hence, the calculated chemical formula of natural copper is $Cu_{0.9167}Fe_{0.0585}Ni_{0.0248}$. It is interesting to point out that the Ni content (2.22–2.25 wt%) in Suizhou natural copper is very close to the Ni content of natural copper ($\sim$1.5–2.0 wt%) in Blansko and Jelica meteorites, and element Ni might exist in solid solution with copper and iron (Olsen 1973; Rubin 1994).

3.4 Phosphate Minerals

Phosphate minerals typically constitute less than 2 vol.% of ordinary chondrites. However, the distribution of phosphates may be heterogeneous within any chondrite. Two phosphate mineral phases, whitlockite (merrillite) and chlorapatite, are identified in chondrites. The relative proportions of the two phosphate minerals vary from chondrite to chondrite, and whitlockite is usually more abundant than chlorapatite.

3.4.1 Whitlockite (Merrillite)

Whitlockite is an anhydrous phosphate mineral. Its composition is essentially $Ca_3(PO_4)_2$, with significant substitutions of Na and Mg for Ca, so that its formula is close to $2.55 CaO \cdot 0.28 MgO \cdot 0.15 Na_2O \cdot 0.02 FeO \cdot P_2O_5$. Phosphates are major host of U, Pu, and REE in ordinary chondrites. Typical U concentrations in whitlockite and chlorapatite are 200 ppb and 3 ppm. In contrast to U, Pu and REE are more enriched in whitlockite than in chlorapatite.

Occurrence
The Suizhou meteorite contains about 1.5 % by volume of whitlockite and <1 % by volume of chlorapatite. Whitlockite in the chondritic rock occurs as single euhedral or semi-euhedral crystals up to 150 µm in grain size. Figures 3.53 and 3.54 demonstrate rather small grains of whitlockite in association with olivine and plagioclase in the Suizhou unmelted chondritic rock. Our microscopic and SEM studies indicate that the whitlockites in this meteorite are usually fractured or heavily fractured and in some cases, even smashed to tiny fragments (Fig. 3.55), but its crystal structure is undisturbed.

Chemical Composition
Our EPMA analysis shows that the Suizhou whitlockite contains 46.62 wt% of CaO, 47.67 wt% of P_2O_5, 3.27 wt% of MgO, and 2.57 wt% of Na_2O (Table 3.14), which are similar to whitlockites in many other ordinary chondrites and achondrites, but distinct from REE-rich whitlockites found in lunar rocks and hydrogen-bearing whitlockites in terrestrial samples (Dowty 1977; Prewitt 1975).

Fig. 3.53 BSE image showing a small whitlockite (*Wht*) grain in the Suizhou meteorite. The upper part of this grain has been damaged during the preparation of the polished section.
Olv = olivine; FeS = troilite; Chr = chromite

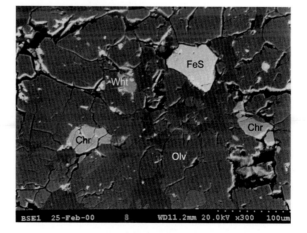

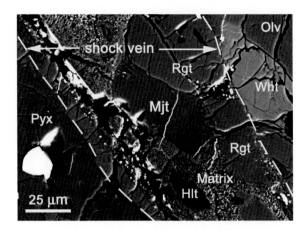

Fig. 3.54 BSE image showing a whitlockite (*Wht*) grain adjacent to a shock melt vein in the Suizhou meteorite. Note sharp boundary between whitlockite and the shock vein. Rgt = ringwoodite; Mjt = majorite; Hlt = lingunite, Olv = olivine, Pyx = pyroxene, Matrix = fine-grained vein matrix (Chen et al. 2004b)

Fig. 3.55 BSE image showing a heavily fractured and smashed whitlockite (*Wht*) grain in the Suizhou meteorite. Olv = olivine, Pyx = pyroxene, Plg = plagioclase, FeNi = FeNi metal

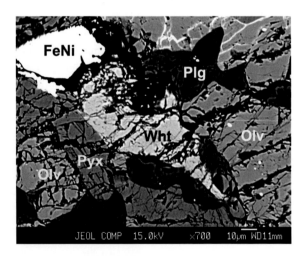

Raman Spectroscopy

Raman spectroscopic investigation indicates that the spectrum of whitlockite in Suizhou consists of intense peaks at 959 and 972 cm^{-1}, less intense peaks at 410, 448, and 1081 cm^{-1}, and weak peaks at 184, 549, 1030, and 1109 cm^{-1} (Fig. 3.56). The two strongest peaks at 959 and 972 cm^{-1} can be assigned to the v1 symmetric stretching vibration of PO$_4$ tetrahedra (Jolliff et al. 1996). The Raman spectrum of this whitlockite is similar to the Raman spectra of whitlockite in the Sixiangkou meteorite (Chen et al. 1995) and synthetic REE-poor whitlockite (Jolliff et al. 1996), where the spectra contain well-resolved strong doublet from 950–976 cm^{-1}.

Table 3.14 Chemical composition of whitlockite in Suizhou (wt%)

Oxides	SZ-Wht	S-2[a]	S-1-2 $_1$[a]	S-1-2 $_2$[a]
TiO_2	0.06	0.03	0.00	0.15
FeO	0.28	0.39	0.11	0.33
MgO	3.27	3.22	3.24	3.35
CaO	46.62	46.36	46.74	46.77
Na_2O	2.57	2.54	2.67	2.50
K_2O	0.03	0.02	0.00	0.08
P_2O_5	47.67	48.10	48.07	46.75
Cr_2O_3	0.03	0.08	0.02	0.00
NiO	0.08	0.16	0.01	0.06
Total	100.61	100.90	100.86	100.09

All values were determined by EPMA in weight %
[a]Data from Zeng (1990)

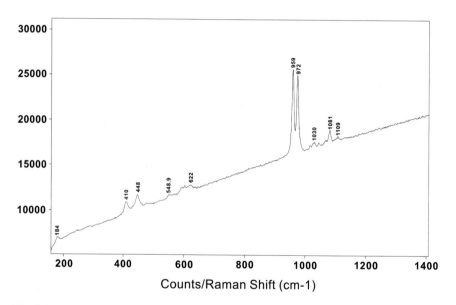

Fig. 3.56 Raman spectrum of whitlockite in the Suizhou

Shock Deformation Features

Whitlockites in the Suizhou meteorite is usually fractured or heavily fractured. Wide cracks filled with molten plagioclase can also be observed in some whitlockite grains (Fig. 3.58). Furthermore, in some cases, the whitlockite grains were smashed to very tiny fragments of irregular shape with grain size ranging from 0.5 to 3–4 μm (Fig. 3.58), but their crystal structure still keeps unchanged.

3.4.2   Chlorapatite

Apatite $(Ca_5(PO_4)_3(F,OH,Cl))$ is among the most important phosphate minerals found in terrestrial rocks, lunar samples, and meteorites (Griffin et al. 1972; Prewitt 1975; Dowty 1977; Bushward 1984; Nach 1984; Rubin 1997; Engvik et al. 2009). Under reflected light, apatite looks like anhydrous phosphate mineral whitlockite, but it is a volatile-bearing phosphate. There are three types of apatites, namely hydroxylapatite $(Ca_3(PO_4)_2(OH))$, fluorapatite $(Ca_3(PO_4)_2F)$, and chlorapatite $(Ca_5(PO_4)_3Cl)$. It is known that apatite occurring in meteorites is mainly in chlorapatite in composition. It contains chlorine and little fluorine, but does not contain hydroxyl. The Suizhou meteorite contains about <1 % by volume of chlorapatite. This mineral was characterized in situ on polished thin sections using scanning electron microscopy, electron microprobe analyses, and Raman spectroscopy.

Occurrence
In the Suizhou chondrite, chlorapatite occurs as tiny irregular grains in the interstices of matrix minerals with grain size ranging from 0.01 to 0.10 mm. Under reflected light or SEM in back-scattered electron mode, chlorapatite is light gray in color. At first appearance, chlorapatite shows no difference to whitlockite. It is revealed that most chlorapatite grains in the Suizhou meteorite are fractured in different level (Figs. 3.57, 3.58, 3.59, and 3.60).

   Figures 3.57 and 3.58 show two less fractured chlorapatite grains of different size, which are surrounded by olivine and plagioclase grains. However, we still can observe some open or closed fractures in these grains. Figure 3.59 demonstrates a moderately shocked larger chlorapatite grain in the Suizhou meteorite. It has a grain size of 60–70 μm in width and 180 μm in length and is surrounded by olivine, pyroxene, and plagioclase. Besides some irregular fractures and cracks, two sets of

Fig. 3.57 BSE image showing an irregular chlorapatite (*Apt*) grain in the Suizhou chondrite.
Ol = olivine,
Plg = plagioclase

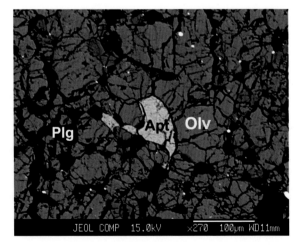

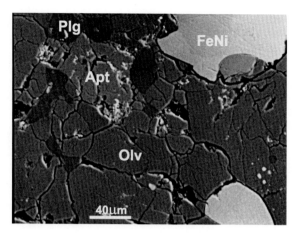

Fig. 3.58 BSE image showing a less fractured chlorapatite (*Apt*) grain in the Suizhou chondrite. The surface of this grain was damaged during the preparation of polished section. Olv = olivine; Plg = plagioclase; FeNi = FeNi metal

Fig. 3.59 BSE image showing a moderately fractured chlorapatite (*Apt*) grain in the Suizhou chondrite. Note the loosely spaced planar fractures and irregular cracks in this grain. Ol = olivine; Pyx = pyroxene, Plg = plagioclase

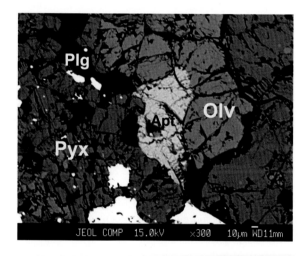

planar fractures are developed in this grain. Figure 3.60 demonstrates a rather heavily shocked large chlorapatite grain of 80–160 μm in width and 200 μm in length. The presence of cracks, planar fractures, and fragmentation indicates that this grain is more heavily damaged by shock compression.

Chemical Composition

Results of the electron microprobe analysis of chlorapatite in the Suizhou meteorite are shown in Table 3.15, giving the following average composition in wt%: CaO— 52.87, P_2O_5—42.41, Cl—5.47, and F—0.58. It should be pointed out that the content of Na_2O in chlorapatite is as low as to 0.216 wt% and that of MgO is only 0.020 wt%. If we compare the composition of chlorapatite with that of whitlockite

Fig. 3.60 BSE image showing a large heavily fractured chlorapatite (*Apt*) grain in the Suizhou chondrite. Note the rather loosely spaced parallel fractures. Olv = olivine; Mas = Maskelynite

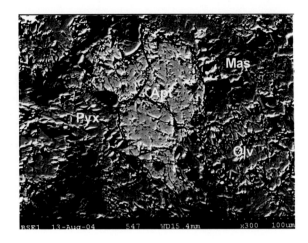

Table 3.15 Chemical composition of chlorapatite in Suizhou meteorite (wt%)

	SZ-A-1	SZ-A-2	SZ-A-3	SZ-A-4	Average
Na_2O	0.22	0.19	0.20	0.25	0.22
K_2O	–	0.01	–	–	–
MgO	–	–	–	0.08	0.02
CaO	52.58	52.69	53.38	52.83	52.87
FeO	–	–	–	0.02	–
P_2O_5	42.49	42.82	42.67	41.64	42.41
F	0.59	0.53	0.62	0.57	0.58
Cl	5. 56	5.73	4.85	5.74	5.47
	101.44	101.97	101.72	101.13	101.57
–O=Cl+F	1.50	1.52	1.36	1.53	1.48
Total	99.94	100.45	100.36	99.60	100.09

in the same Suizhou meteorite (Table 3.15), we can see that the Suizhou chlorapatite contains more CaO, less P_2O_5, and much less Na_2O and MgO than those in the Suizhou whitlockite.

Raman Spectroscopy

The Raman spectrum of chlorapatite in the Suizhou meteorite shows an intense peak at 960 cm^{-1}, three less intense peaks at 427, 589, and 1038 cm^{-1}, and two weak peaks at 612 and 1077 cm^{-1} (Fig. 3.61). This spectrum is quite similar to that of terrestrial chlorapatite collected from Ødegården verk, Norway.

Our measurements show that the Raman spectrum of Norwegian chlorapatite is also characterized by an intense peak at 961 cm^{-1}, three less intense peaks at 428, 587, and 1034 cm^{-1}, and two weak peaks at 612 and 1078 cm^{-1}. Engvik et al. (2009) reported that chlorapatite from Ødegården verk contains 52.81–54.79 wt% of CaO, 40.15–41.14 wt% of P_2O_5, 5.07–6.81 wt% of Cl, 0.00–0.02 wt% of F, and 0.26–0.35 wt% of Na_2O. Hence, it is reasonable to assume that the extraterrestrial

Fig. 3.61 Raman spectra of chlorapatite in the Suizhou meteorite

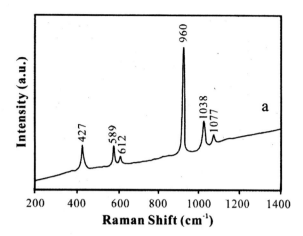

chlorapatite of the Suizhou meteorite and the terrestrial chlorapatite of Ødegården verk are quite similar both in chemical composition and crystal structure.

Shock Deformation Features

Fractures and cracks are rather common shock-induced deformation features for chlorapatite in the Suizhou unmelted chondritic rock. However, some large chlorapatite grains were heavily broken, and up to four sets of planar fractures can be recognized in them (Fig. 3.62). The spacing between two adjacent fractures is in the range from 2 to 5 μm. Furthermore, some chlorapatite grains were heavily broken or smashed to many small fragments of irregular shape (Fig. 3.63), but our Raman spectroscopic study on these grains shows that the crystal structure of chlorapatite is still intact.

Fig. 3.62 BSE image showing a large heavily fractured and fragmented chlorapatite (*Apt*) grain in the Suizhou unmelted chondritic rock. Note the four sets of planar fractures in this grain. Olv = olivine; Plg = plagioclase

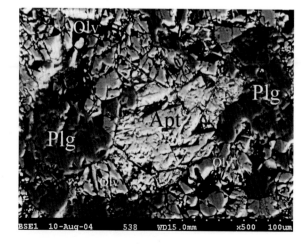

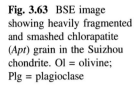

Fig. 3.63 BSE image showing heavily fragmented and smashed chlorapatite (*Apt*) grain in the Suizhou chondrite. Ol = olivine; Plg = plagioclase

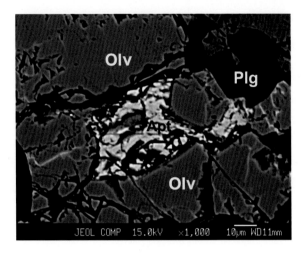

3.5   Summary

1. The rock-forming minerals of the Suizhou meteorite are olivine, pyroxene, and plagioclase. They make about 88 % of the meteorite by volume. Olivine and pyroxene show mosaic shock feature, while most of plagioclase grains have transformed to maskelynite.
2. FeNi metal (kamacite and taenite) and troilite are minor minerals in the Suizhou meteorite. They are about 10 % of the meteorite by volume. Some small rounded FeNi metal grains formed from shock-induced FeNi metal or condensed from shock-produced FeNi vapor were observed. Fine-scale intergrowth of the minor 4C monoclinic pyrrhotite and the primary 2C troilite superstructures was revealed in the Suizhou troilite.
3. Chromite is a common accessory mineral in the Suizhou meteorite. It makes ~1.4 % of the meteorite by volume. Chromite occurs either in the form of single coarse grains or as tiny fragments embedded in molten plagioclase. The exolution chromite is extremely rare.
4. Two phosphate mineral phases, whitlockite (merrillite) and chlorapatite, are identified in the Suizhou meteorite. Whitlockite is more abundant (1.5 % by volume) than chlorapatite (<1 % by volume) in this meteorite. Both phosphate minerals were heavily fractured by shock.
5. Ilmenite and natural copper are very rare opaque minerals in the Suizhou meteorite. Only two small ilmenite grains and very few natural copper grains were found on polished thin sections of the Suizhou meteorite.

References

Bennett CEG, Graham J, Thornber MR (1972) New observations on natural pyrrhotites part I. Mineragraphic techniques. Am Mineral 57:445–462

Bertaut EF (1953) Contribution à l'étude des structure lacunaires: La pyrrrhotine. Acta Crystal 6:557–561

Besrest FK, Collin G (1990a) Structural aspects of the α transition in stoichiometric FeS: identification of the high-temperature phase. J Solid State Chem 84:194–210

Besrest FK, Collin G (1990b) II. Structural aspects of the α transition in off-stoichiometric $Fe_{1-x}S$ crystals. J Solid State Chem 84:211–225

Brearly AJ, Jones RH (1998) Chondritic materials. In: Planetary materials. Review in Mineralogy, Virginia, vol 36, pp 3.1–3.389

Buchwald VF (1984) Phosphate minerals in meteorites and lunar rocks [C]. In: Nriagu JO, Moore PB (eds) Phosphate minerals. Springer, Berlin, pp 199–214

Carpenter RH, Desborough GA (1964) Range in solid solution and structure of naturally occurring troilite and pyrrhotite. Am Mineral 49:1350–1365

Chen JZ (1990) A transmission electron microscopic study of the Suizhou meteorite (in Chinese). A synthetical study of Suizhou meteorite. Publishing House of the China University of Geosciences, Wuhan, pp 57–62

Chen M, Wopenka B, Xie XD, El Goresy A (1995) A new high-pressure polymorph of chlorapatite in the shocked chondrite Sixiangkou (L6). Lunar Planet Sci 26:237–238

Chen M, El Goresy A (2000) The nature of maskelynite in shocked meteorites: not diaplectic glass but a glass quenched from shock-induced dense melt at high-pressures. Earth Planet Sci Lett 179:489–502

Chen M, Shu JF, Xie XD et al (2003) Natural $CaTi_2O_4$-structured $FeCr_2O_4$ polymorph in the Suizhou meteorite and its significance in mantle mineralogy. Geochim Cosmochim Acta 67 (20):3937–3942

Chen M, Xie XD, El Goresy A (2004a) A shock-produced (Mg, Fe)SiO3 glass in the Suizhou meteorite. Meteor Planet Sci 39:1797–1808

Chen M, El Goresy A, Gillet P (2004b) Ringwoodite lamellae in olivine: clue to olivine-ringwoodite phase transition mechanisms in shocked meteorites and in subducting slabs. Proc Natl Acad Sci U S A 101:15033–15037

Dódony I, Pósfai M (1990) Pyrrhotite superstructures. Part II: a TEM study of 4C and 5C structures. Eur J Mineral 2:529–535

Dowty E (1977) Phosphate in Angra Dos Reis: structure and composition of the $Ca_3(PO_4)_2$ minerals. Earth Planet Sci Lett 35:347–351

El Goresy A (1976) Opaque oxide minerals in meteorites. Rev Mineral 3:47–72

Engelhardt WV, Arndt J, Stöffler D et al (1967) Diaplektische Gläser in den Breccien des Ries von Nördlingen als Anzeihen für Stoß wellenmetamorphose. Contrib Mineral Petrol 15:91–100

Engvik AK, Golla-Schindler U, Berndt J et al (2009) Intragranular replacement of chlorapatite by hydroxy-fluor-apatite. Lithos 112:236–246

Ericsson T, Amcoff Ö, Nordblad P (1997) Superstructure formation and magnetism of synthetic selenian pyrrhotites of $Fe_7(S_{1-y}Se_y)_8$, $y \leq 1$ composition. Eur J Mineral 9:1131–1146

Evans HT Jr (1970) Lunar troilite: crystallography. Science 167:621–623

Fleet ME (2006) Phase equilibria at high temperatures. Rev Mineral Geochem 61:365–419

Ge YY, Ren YX, Hua X et al (1990) Investigations on Mössbaur effects of the Suizhou meteorite. In: A synthetical study of Suizhou meteorite. Publishing House of the China University of Geosciences, Wuhan, pp 101–107 (in Chinese)

Griffin WL, Åmli R, Heier KS (1972) Whitlockite and apatite from lunar rock 14310 and from ÖdegÅrden, Norway. Earth Planet Sci Lett 15:53–68

Groves DI, Ford RJ (1963) Mineralogical notes: note on the measurement of pyrrhotite composition in the presence of both hexagonal and monoclinic phases. Am Mineral 48:911–913

Hou W, Ouyang ZY, Li ZH, et al (1990) Suizhou, the new fall of a L-group chondrite. In: A synthetical study of Suizhou meteorite. Publishing House of the China University of Geosciences, Wuhan, pp 30–32 (in Chinese)

Jolliff BL, Freeman JJ, Wopenka B (1996) Structural comparison of lunar, terrestrial, and synthetic whitlockite using laser Raman microprobe spectroscopy. Lunar Planet Sci 27:613–614

Kontny A, de Wall H, Sharp TG et al (2000) Mineralogy and magnetic behavior of pyrrhotite from a 260 °C section at the KTB drilling site, Germany. Am Mineral 85:1416–1427

Kuebler KE, McSween HY Jr, Carlson WD et al (1999) Sizes and masses of chondrules and metal-troilite grains in ordinary chondrites: possible implications for nebular sorting. Icarus 141:96–106

Langenhorst F, Poirier JP (2000) 'Ecologitic' minerals in a shocked basaltic meteorite. Earth Planet Sci Lett 176:259–265

Langenhorst F, Stöffler D, Keil K (1991) Shock metamorphism of the Zagami achondrite. Lunar Planet Sci 22:779–780

Makovicky E (2006) Crystal structures of sulfides and other chalcogenides. Rev Mineral Geochem 61:7–125

McCoy TJ, Taylor GJ, Keil K (1992) Zagami: product of two-stage magmatic history. Geochim Cosmochim Acat 56:3571–3582

Milton DJ, Carli PS (1963) Maskelynite formation by explosive shock. Science 140:670–671

Morimoto N, Nakazawa H, Nishiguchi K et al (1970) Pyrrhotites: stoichiometric compounds with composition $Fe_{n-1}S_n$ (n < 8). Science 168:964–966

Morimoto N, Gyobu A, Tsukuma K, Koto K (1975) Superstructure and nonstoichiometry of intermediate pyrrhotite. Am Mineral 60:240–248

Nakazawa H, Morimoto N, Watanabe E (1975) Direct observation of metal vacancies by high-resolution electron microscopy. Part I: 4C type pyrrhotite (Fe_7S_8). Am Mineral 60:359–366

Nash WP (1984) Phosphate minerals in terrestrial igneous and metamorphic rocks. In: Nriagu JO, Moore PB (eds) Phosphate minerals. Springer, Berlin, pp 215–241

Ohfuji H, Sata N, Kobayashi H et al (2007) A new high-pressure and high-temperature polymorph of FeS. Phys Chem Mineral 34:335–343

Olsen EJ (1973) Copper-nickel alloy in the Blansko chondrite. Meteoritics 8:259–261

Ostertag R (1983) Shock experiments on feldspar crystals. J Geophys Res 88:B364–B376

Pósfai M, Sharp TG, Kontny A (2000) Pyrrhotite varieties from the 9.1 km deep borehole of the KTB project. Am Mineral 85:1406–1415

Prewitt CT (1975) Meteoritic and lunar whitlockites. Lunar Planet Sci 6:647–648

Prewitt CT, Downs RT (1998) High-pressure crystal chemistry. Rev Mineral 37:287–317

Putnis A (1974) Electron-optical observations on the α-transformation in troilite. Science 186:439–440

Putnis A (1975) Observations on coexisting pyrrhotite phases by transmission electron microscopy. Contrib Mineral Petrol 52:307–313

Ramdohr P (1967) Chromite and chromite chondrules in meteorites—I. Geochim Cosmochim Acta 31:1961–1967

Ramdohr P (1973) The opaque minerals in stony meteorites. Elsevier Press, Amsterdam, p 245

Rubin AE (1994) Euhedral tetrataenite in the Jelica meteorite. Mineral Mag 58:215–221

Rubin AE (1997) Mineralogy of meteorite groups. Meteor Planet Sci 32:231–247

Rubin AE (2003) Chromite-plagioclase assemblages as a new shock indicator: implications for the shock and thermal histories of ordinary chondrites. Geochim Cosmochim Acta 67:2695–2709

Schmitt RT (2000) Shock experiments with the H6 chondrite Kernouve: Pressure calibration of microscopic shock effects. Meteor Planet Sci 35:545–560

Shen SY, Zhuang XL (1990) A study of the opaque minerals and structural characteristics of the Suizhou meteorite. In: A synthetical study of Suizhou meteorite. Publishing House of the China University of Geosciences, Wuhan, pp 40–52 (in Chinese)

Skála R, Císařová I (2005) Crystal structure of troilite from chondrites Etter and Georgetown. Lunar Planet Sci 36:1284

Skála R, Císařová I, Drábek M (2006) Inversion twinning in troilite. Am Mineral 91:917–921

Smith BA, Goldstein JI (1977) The metallic microstructures and thermal histories of severely reheated chondrites. Geochim Cosmochim Acta 41:1061–1072

Snetsinger KG, Keil K, Bunch TE (1967) Chromite from "equilibrated "chondrites. Am Mineral 52:1322–1331

Stöffler D, Keil K, Scott ED (1991) Shock metamorphism of ordinary chondrites. Geochim Cosmochim Acta 55:3854–3867

Tokonami M, Nishiguchi K, Morimoto N (1972) Crystal structure of a monoclinic pyrrhotite (Fe_7S_8). Am Mineral 57:1066–1080

Töpel-Schadt J, Müller WF (1982) Transmission electron microscopy on meteoritic troilite. Phys Chem Mineral 8:175–179

Tschermak G (1872) Die Meteoriten von Shegotty und Gopalpur, Sitzber. Akad. Wiss. Wien Math.-Natrurwiss. Kl. Abt. I 88:347–271

Wang DD (1993) An Introduction to Chinese meteorites. Science Press, Beijing, pp 101–106 (in Chinese)

Wang DD, Li ZH (1990) A study of rare gas dating of the Suizhou meteorite. In: A synthetical study of Suizhou meteorite. Publishing House of the China University of Geosciences, Wuhan, pp 125–128 (in Chinese)

Wang HP, Salveson I (2005) A review on the mineral chemistry of the non-stoichiometric iron sulphide, $Fe_{1-x}S$ ($0 \leq x \leq 0.125$): polymorphs, phase relations and transitions, electronic and magnetic structures. Phase Transit 78:547–567

Wang RJ, Qiu JR, Shen JC (1990) A preliminary study of the Suizhou meteorite. In: A synthetical study of Suizhou meteorite. Publishing House of the China University of Geosciences, Wuhan, pp 12–19 (in Chinese)

Weber I, Semenenko VP, Stephan T et al (2006) TEM studies and the shock history of a "mysterite" inclusion from the Krymka LL chondrite. Meteorit Planet Sci 41:571–580

Xie XD, Chen M, Wang DQ (2001a) Shock-related mineralogical features and P-T history of the Suizhou L6 chondrite. Eur J Miner 13(6):1177–1190

Xie XD, Chen M, Wang DQ et al (2001b) $NaAlSi_3O_8$-hollandite and other high-pressure minerals in the shock melt veins of the Suizhou L6 chondrite. Chin Sci Bull 46(13):1121–1126

Xie XD, Chen M, Wang DQ et al (2006) Melting and vitrification of plagioclase under dynamic high pressures. Acta Petrologica Sin 22:503–509 (In Chinese with an English abstract)

Xie XD, Chen M, Wang CW (2011) Occurrence and mineral chemistry of chromite and xieite in the Suizhou L6 chondrite. Sci Chin Earth Sci 54:1–13

Zeng GC (1990) Common transparent minerals and chemico-petrological type of the Suizhou meteorite. In: A synthetical study of Suizhou meteorite. Publishing House of the China University of Geosciences, Wuhan, pp 32–40 (in Chinese)

Zhang K, Wang JB, Wang RH (2006) Electron microscopic studies of the mineral phases in the Suizhou meteorite. J Chin Electr Microsc Soc 25 (Suppl):354–355 (in Chinese)

Zhang K, Zheng H, Wang JB, Wang RH (2008) Transmission electron microscopy on iron sulfide varieties from the Suizhou meteorite. Phys Chem Minerals 35:425–432

Chapter 4
Distinct Morphological and Petrological Features of the Suizhou Shock Veins

4.1 General Remarks

Heavily shocked chondritic meteorites, for example those in the L and H groups, frequently contain shock-produced melt veins that have the same chemical composition as the chondritic host rock. Shock-induced melts occur in different textural settings and geometries in all groups of ordinary chondrites (e.g., Dodd and Jarosewich 1979; Dodd et al. 1982; Ashworth 1985). Stöffler et al. (1991) proposed to distinguish the following types of shock-induced localized melting: (1) thin opaque melt veins, consisting of "mixed" melt products of (rarely) glassy, mostly aphanitic, polycrystalline material formed by in situ melting of various mineral constituents; (2) melt pockets and interconnected irregular melt veins with aphanitic matrix and irregular boundaries to the host; (3) melt dikes which consist of crystalline, mixed melt products, with fine-grained polycrystalline matrix and discordant boundaries to the host; and (4) sulfide and metal deposits injected into fractures in olivine and pyroxene that is frequently referred to as "shock blackening" in the literature.

It is revealed by Stöffler et al. (1991) that most of the shock melt veins in chondrites occur in irregular and curved forms and hyphazardly traverse the chondritic meteorites. Sometimes, one vein branches into two or more veins later feeding back into the main vein or continueing into different directions at oblique angles. For instance, the Pervomaisky (L6), Pinto Mountains (L6), and Cangas de Onis (H5) chondrites are intensely penetrated by branched black melt veins of variable width (tens of μm—3 mm) (Semenenko and Golovka 1994; Stöffler et al. 1991). Pervasive networks of shock veins were observed in the Chantonnay, Pampa del Infierno, Sixiangkou and some other L6 chondrites (Dodd et al. 1982; Boctor et al. 1982; Chen et al. 1996a), as well as in the Yanzhuang H6 chondrite (Xie et al. 1991). For most vein-bearing chondrites, vein width ranges from 0.1 to 10 mm. The thickest shock vein observed by us measures 15 mm in the Yanzhuang H6 chondrite (Xie et al. 1991), but the average width for shock veins in chondrites is estimated at 0.5–1 mm. Only shock veins of L group chondrites were reported as

© Springer-Verlag Berlin Heidelberg and Guangdong Science
& Technology Press Co., Ltd. 2016
X. Xie and M. Chen, *Suizhou Meteorite: Mineralogy and Shock Metamorphism*,
Springer Geochemistry/Mineralogy, DOI 10.1007/978-3-662-48479-1_4

occurrences of high-pressure phases such as ringwoodite, wadsleyite, majorite and others (e.g., Binns et al. 1969; Coleman 1977; Madon and Poirier 1979; Tomioka and Fujino 1997). Rubin (1985) pointed out that ringwoodite and majorite were found in shock veins of seven L6 chondrites. More recently, high-pressure minerals were also found in melt veins of many other L6 chondrites, such as Sixiangkou, Peace River, Mbale, Acfer-040, Suizhou, Y-791384, Y-74445, Umbarger and S-98222 (Chen et al. 1996a, b, 1998b; Sharp et al. 1997a, b; Xie et al. 2001a, b, c, 2002; Kimura et al. 2000; Ozawa et al. 2009). Tiny ringwoodite grains were identified by Raman spectroscopy in the heavily shock-metamorphosed mosaic olivine of the Yanzhuang H6 chondrite (Chen and Xie 1993a, b). Ringwoodite, majorite and lingunite ($NaAlSi_3O_8$-hollandite) were also observed in the Antarctic H-group Y-95267 (H6) and Y-75100 (H6) chondrites (Kimura et al. 2000, 2003).

The Suizhou meteorite was evaluated as a strongly shockd (S5) meteorite due to the presence of mosaic olivine and transformation of plagioclase to maskelynite (Xie et al. 2001a; Chen et al. 2008). In comparison to other L6 chondrites containing shock veins up to several millimeters in width (Chen et al. 1996a; Price et al. 1979; Rubin 1985), the Suizhou meteorite contains only a few very thin black melt veins ranging from 0.02 to 0.09 mm in width. Morphologically, the thin Suizhou melt veins look like fine parallel fractures traversing the chondritic meteorite, but they are distinct straight solid melt veins. Petrologically, they are chondritic in composition and consist of two distinct high-pressure mineral assemblages: (a) coarse-grained polycrystalline ringwoodite, majorite, and lingunite that are formed through the solid state transformations of olivine, low-calcium pyroxene and plagioclase, respectively; and (b) fine-grained majorite-pyrope garnet in solid solution + magnesiowüstite that crystallized at high pressures and temperatures from shock-induced dense chondritic melt (Xie et al. 2001a). Fine-grained garnet + magnesiowüstite together with FeNi-metal and troilite intergrowths make up the matrix of the vein. It is interesting that our transmission electron microscopic study revealed that the Suizhou vein matrix consists of fully crystalline material with no glassy material remaining. In this chapter we describe the distinct morphological and petrological features of the shock melt veins in the Suizihou L6 chondrite.

4.2 Morphological Features of the Suizhou Melt Veins

4.2.1 The Simplest Melt Veins

Comparing with the irregular-shaped, curved or branched melt veins, and some of the complex or network-like melt veins in many L and H group chondrites, the melt veins in the Suizhou meteorite are the simplest (Xie et al. 2001a).

For instance, Coleman (1977) described a network of dark veins on sawn surface of the Catherwood L6 chondrite. Dodd et al. (1982) reported the first occurrence of complex veins in the Chantonnay L6 chondrite, in which an early formed vein

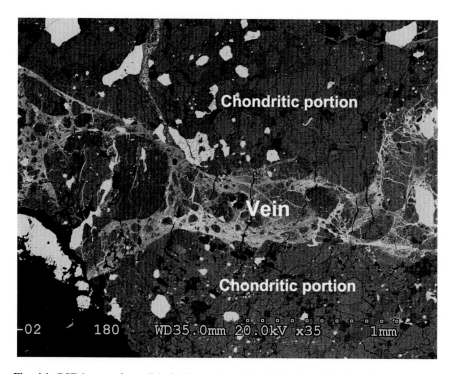

Fig. 4.1 BSE image of a polished thin section of the Sixiangkou L6 chondrite showing the occurrence of a melt vein. Note the curved and jagged boundaries between the vein and the chondritic portion and a branched veinlet (Xie et al. 2011a)

containing abundant chondritic xenoliths is intersected by a late vein composed of a fine-grained matrix. Boctor et al. (1982) found complex vein geometries in the form of networks in the Pampa del Infierno chondrite. The well known Peace River L6 meteorite also consists of a severely deformed chondrite mass and a network of melt veins up to 3 mm in width (Chen et al. 1998b). A network of dark veins was also observed in the Sixiangkou L6 chondrite that contains a number of curved and branched veins intersecting the chondritic host (Fig. 4.1). More recently, we identified shock-produced complex veins, including earlier and later veins, in this Sixiangkou chondrite. The early veins are intersected by the later veins and consist of both coarse-grained aggregates of ringwoodite, majorite and lingunite, and fragments of olivine, pyroxene, plagioclase, metal and troilite, as well as a fine-grained matrix of garnet, ringwoodite, metal and troilte. The later veins mainly consist of a fine-grained matrix of garnet, magnesiuwüstite, metal and troilite, with a small amount of coarse-grained aggregates (Chen and Xie 2008). The Yanzhuang H6 chondrite was also found to consist of a chondritic matrix and network of melt veins ranging in width from 0.1 to 15 mm and melt pockets up to 30 mm in size (Xie et al. 1991; Chen et al. 1998a). A network of dark shock veins penetrating the entire chondritic rock is also observed in the Yanzhuang chondrite studied here (Fig. 4.2).

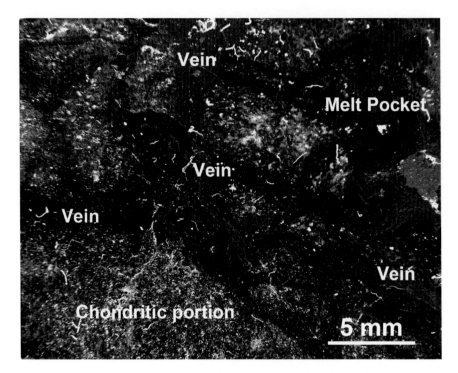

Fig. 4.2 Photograph of sawn surface of the Yanzhuang H6 chondrite showing a network of dark melt veins. Note a melt pocket in the *upper right part* of the photograph (Xie et al. 2011a)

The morphology of Suizhou melt veins is remarkably different from that in most chondritic meteorites. Figure 4.3 shows four thin melt veins in a large Suizhou fragment of 1.434 kg in weight. All of them are single veins. Among them, three are parallel to each other and extend in the same direction, and the fourth is an oblique vein joining the middle one of the three others at an angle of about 45°. Neither intersection of veins nor vein network were observed in the fragment (Xie et al. 2001a). Hence, we assume that the melt veins in the Suizhou L6 chondrite are the simplest single veins in comparison with those in other vein-bearing chondrites.

4.2.2 The Straightest Melt Veins

The melt veins in the Suizhou meteorite are also the straightest compared with the curved or irregular melt veins in many other chondrites (Xie et al. 2001a).

It is known that melt veins in most chondritic meteorites occur in curved form and show irregular vein boundaries to the host chondrite. For instance, the Tenham, Coorara, Catherwood, and Coolaman chondrites contain irregular or complex melt veins traversing the chondritic meteorites (Steel and Smith 1978). The networks of

Fig. 4.3 Photograph of a fragment of the Suizhou meteorite showing three thin parallel and one oblique melt vein (*black lines*) (Xie et al. 2011a)

dark veins observed in the Peace River, Pampa del Infierno and Sixiangkou L6 chondrites, as well as in the Yanzhuang H6 chondrite, also display irregular vein edges. On the other hand, melt veins in the Suizhou meteorite are quite regular and show very straight vein boundaries. Figures 4.4 and 4.5 show some very straight melt veins in the Suizhou meteorite that have knife-sharp boundaries to the host chondrite on both sides of the veins. Furthermore, we observed that molten troilite often occurs in the form of straight thin strings of more than 100 μm length and less than 3 μm width along the boundaries of a vein bordering the unmelted part of the meteorite (Fig. 4.5).

It should be pointed out that neither melt dikes nor melt pockets were observed in the Suizhou meteorite. Therefore, it is reasonable to assume that the shock-produced melt veins in the Suizhou chondrite are the straightest in comparison with those in most vein-bearing chondrites.

4.2.3 The Thinnest Melt Veins

The melt veins in the Suizhou meteorite are the thinnest among all other shock-vein-bearing meteorites (Xie et al. 2001a).

It has been reported that shock melt veins in chondritic meteorites are about 1 mm in width, such as those in the Tenham, Coorara, Catherwood, and Coolaman L6 chondrites (Steel and Smith 1978). Some variations in vein thickness were observed in different meteorites, but it is hard to find veins thinner than 0.1 mm in

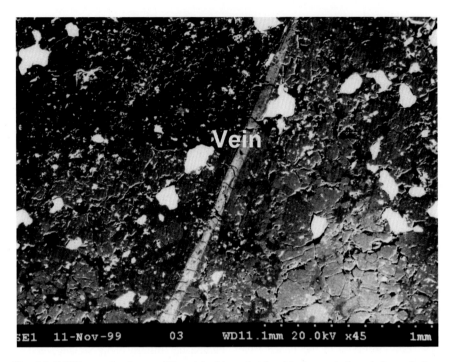

Fig. 4.4 BSE image showing a solid straight thin melt vein with sharp vein boundaries to the host chondrite (Xie et al. 2011a)

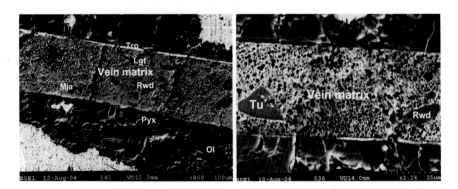

Fig. 4.5 BSE images showing straight melt veins with sharp vein boundaries to the host chondrite. Note the two distinct lithological assemblages: fine-grained vein matrix and coarse-grained mineral fragments. Ol = olivine, Pyx = pyroxene, Rwd = ringwoodite, Mja = majorite, Lgt = lingunite, Tro = troilite, Tu = tuite, Matrix = vein matrix (Xie et al. 2011a)

most of these cases. For instance, the shock veins in Catherwood meteorite range from 0.1 to 10 mm in apparent width, and the average true width is estimated to be 0.5 mm (Coleman 1977). The width of Sixiangkou shock veins ranges from 10 μm to 10 mm. Besides some simple veins with thicknesses less than 200 μm, there are complex veins with thickness ranging from 200 μm to 10 mm (Chen and Xie 2008). However, almost all the melt veins in the Suizhou chondrite are thinner than 0.1 mm in width, namely, from 0.05 to 0.09 mm, and we have observed only one case where the vein width locally reached 0.2 mm. Figures 4.3 and 4.4 demonstrate that the shock veins traversing the whole body of a Suizhou fragment are very thin (on average, <0.1 mm in width) and the vein width remains rather uniform.

4.3 Petrological Features of the Suizhou Shock Veins

4.3.1 Black Fracture-like Features Are Solid Melt Veins

It is interesting to point out that at first glance, the Suizhou black veins look like fine fractures traversing the chondritic body of the meteorite. However, they are single and straight solid melt veins (Fig. 4.4). They are chondritic in mineralogical composition and consist of fine-grained vein matrix and comparatively coarse-grained mineral fragments that are randomly distributed in the matrix (Fig. 4.5).

Our microscopic study revealed that the fine-grained matrix of the Suizhou shock veins makes up about 80–90 % of the veins by volume, and the remaining 10–20 % of the veins by volume consist of coarse-grained mineral fragments (Xie et al. 2001a, b, c). With the SEM, the vein matrix is revealed to be composed of numerous tiny euhedral crystals of garnet with 1–5 μm in size which indicates that they crystallized at high-pressure and high-temperature from dense chondritic melt (Fig. 4.6). Tiny irregular grains of magnesiowüstite and ringwoodite are located in the interstices between garnet crystals. Eutectic intergrowths of FeNi-metal and troilite that crystallized during meteorite cooling also occur in the matrix of the veins (Fig. 4.6). On the other hand, the coarse-grained clast assemblage comprises different high-pressure mineral phases that were formed through solid state transformation of silicate, phosphate and oxide minerals (Xie et al. 2001a, b, 2003, 2005a, 2011a; Chen et al. 2003a, b). This demonstrates that these melt veins represent shock veins.

4.3.2 Fully Crystalline Vein Matrix with no Glassy Material Remaining

Stöffler et al. (1991) reported that the shock veins in many chondritic meteorites rarely consist of glassy, and mostly of aphanitic, polycrystalline material formed from in situ melted mineral constituents. Our microscopic, SEM and TEM studies on the mineral composition of shock veins in Yanzhuang H6 and Sixiangkou L6 chondrites have indicated that glassy material can always be observed in veins with width of several millimeters (Chen et al. 1998a; Chen and Xie 2008). However, our

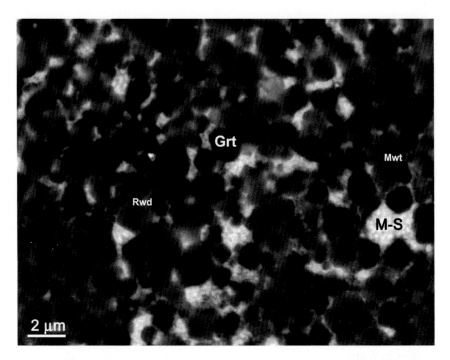

Fig. 4.6 BSE image showing the mineral constituents of the Suizhou vein matrix. Grt = majorite-pyrope garnet in solid solution. Mwt = magnesiowüstite; Rwd = ringwoodite; M-S = metal-troilite intergrowth (Xie et al. 2011a)

SEM and TEM studies on Suizhou vein matrix revealed that besides the metal-troilite eutectic intergrowths that solidified at the last stage of vein formation, the matrix is composed of three fine-grained high-pressure mineral phases with good crystallinity, no glassy material was detected (Xie et al. 2001a). The fine-grained crystalline high-pressure minerals are:

1. Equant idiamorphic majorite-pyrope garnet crystals are the first product of crystallization from the shock-produced dense silicate melt (Fig. 4.7a, b, c). Mocroprobe analyses showed that this garnet mineral contains a much higher content of Al_2O_3 (3.51 wt%) than that in Suizhou pyroxene (0.16 wt%) and relatively coarse-grained majorite (0.16 wt%). The synchrotron radiation X-ray diffraction (SRXRD) patterns of this phase show 16 diffraction reflections. The strongest lines are at 2.572 Å (intensity 100), 2.880 (70), 1.539 (70), 2.453 (50), and 1.597 (40). These data are identical to those for standard majorite in JCPDS No. 25-0843, and our data can be considered as the characteristic X-ray diffraction patterns for the liquidus majorite-pyrope garnet. It is also worth to mention that the diffraction peaks obtained for majorite-pyrope garnet are sharp enough to indicate good crystallinity of this mineral.

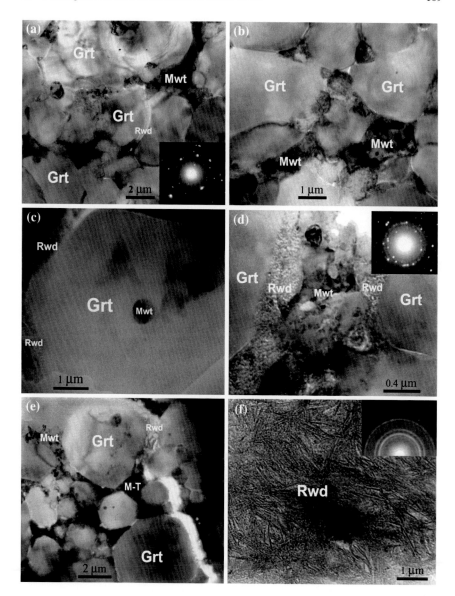

Fig. 4.7 Transmission electron microscopy bright field images **a–f** showing the fine-grained high-pressure minerals in Suizhou shock vein matrix. Grt = majorite-pyrope garnet in solid solution, Mwt = magneisiowüstite, Rwd = ringwoodite. Insets are the SAED patterns of vein minerals: Diffraction spots of magneisiowüstite in (**a**); diffraction spots of magneisiowüstite and diffraction rings of fine-granular ringwoodite in (**d**); and diffraction rings of fiber-like ringwoodite in (**f**) (Xie et al. 2011a)

2. Irregular polycrystalline grains of magnesiowüstite, (Mg, Fe)O, fill the interstices between garnet crystals (Fig. 4.7a–e). Their SAED patterns show sharp diffraction spots of magnesiowüstite (insets of Fig. 4.7a–d), and the calculated strong diffraction lines are: 2.120 (100), 1.492 (60), and 2.439 (40) Å, which can be compared with those of periclase (MgO) in JCPDS No. 4-289. The slight increase of d-values is due to the presence of FeO in its composition.

3. Polycrystalline ringwoodite occurs in the form of fine-granular (Fig. 4.7c, d), blocky (Fig. 4.7e), or fiber-like (Fig. 4.7f) narrow bands between garnet crystals or between garnet and magnesiowüstite grains. Their SAED patterns show only rather sharp concentric diffraction rings. The calculated strongest 4 diffraction lines are at 2.446 (100), 2.028 (70), 1.434 (60), and 2.872 (30) Å, which is identical to the reflections of ringwoodite in JCPDS No. 21-1258.

4.3.3 Abundant Coarse-Grained High-Pressure Minerals in Veins

Besides the three fine-grained high-pressure minerals occurring in matrix, the Suizhou shock veins also contain more abundant coarse-grained high-pressure mineral species, in comparison with all other vein-bearing chondrites, and almost all rock-forming and accessory minerals in the Suizhou shock veins have been transformed to their high-pressure polymorphs (Xie et al. 2001a), and no fragments of the precursor minerals remain in the veinlets (Table 4.1).

Rubin (1985) pointed out that only two high-pressure minerals, ringwoodite and majorite, were found in the shock veins of seven L6 chondrites. In recent years more and more high-pressure phases have been identified in some shocked chondrites (Chen et al. 1996a, 2003b, 2008; Sharp et al. 1997a, b; Tomioka and Fujino 1997; Gillet et al. 2000; Xie et al. 2001a, 2002, 2003; Kimura et al. 2000, 2001; Xie and Sharp 2007). Table 4.1 summarizes the high-pressure phases identified in shock melt veins of different L and H group chondrites. Among 28 shock-vein-bearing chondrites, 17 contain only 1–3 high-pressure phases, 8 meteorites contain 4–6 high-pressure phases, 2 meteorite contains 7 high-pressure phases, and only the Suizhou meteorite contains 10 high-pressure phases. Hence, we assume that the Suizhou chondrite is the most abundant high-pressure-mineral bearing meteorite, although the shock veins in Suizhou are extremely thin.

Our SEM, TEM and Raman spectroscopic studies revealed that besides the above mentioned fine-grained assemblage, crystallized from shock melt, and the FeNi metal and troilite (FeS) which were molten and occur as fine eutectic FeNi–FeS intergrowths in the interstices between fine-grained high-pressure minerals, all coarse-grained rock-forming silicate minerals in the Suizhou shock veins, such as olivine, pyroxene and plagioclase, and almost all accessory phosphate and oxide minerals, such as whitlockite, chlor-apatite and chromite, have been transformed to their high-pressure polymorphs. The only missing accessory mineral in veins is the

Table 4.1 Shock-produced high-pressure minerals in the shock melt veins of ordinary chondrites

Meteorite	Group	High-pressure minerals in veins[a]	Reference
Suizhou	L6	Rwd, Mjt, Lgt, Akm, Pvt, Mwt, Grt, Tut, Xie, CF-phase	Xie et al. (2001a, 2003), Chen et al. (2003a)[b]
Sixiangkou	L6	Rwd, Wds, Mjt, Lgt, Mw, Grt, A-phase	Chen et al. (1995, 1996a), Gillet et al. (2000)
Y-791384	L6	Rwd, Wds, Mjt, Akm, Lgt, Jdt, Grt,	Kimura et al. (2000), Ohtani et al. (2006)
Tenham	L6	Rwd, Mjt,Ca-Mjt, Lgt, Akm, Pvt	Binns et al. (1969), Xie et al. (2006), Xie and Sharp (2007)
GRV-052082 052082052082 052082 052082052082 052082	L6	Rwd, Mja, Lgt, Akm, Grt, Jdt	Feng (2011)
GRV 052049	L5	Rwd, Mja, Lgt, Grt, Tut, Jdt	Feng (2011)
Y-74445	L6	Rwd, Wds, Mjt, Akm, Lgt, Jd	Kimura et al. (1999, 2000), Ozawa et al. (2009)
Umbarger	L6	Rwd, Akt, Lgt, Fe-Sp, Stv	Xie et al. (2002)
Y-95267	H6	Rwd, Wds, Mjt, Akm, Lgt	Kimura et al. (2003)
A-78003	L6	Rwd, Wds, Mjt, Grt, Jdt	Ohtani et al. (2006)
Y-75100	H6	Wds, Mjt, Lgt, Jdt	Kimura et al. (2000)
Peace River	L6	Rwd, Wds, Grt	Sharp et al. (1996); Chen et al. (1998b)
Mbale	L6	Rwd, Wds, Grt	Sharp et al. (1996), Chen et al. (1998b)
Acfer 040	L5/6	Rwd, Akm, Pvt	Sharp et al. (1997a, b)
Yanzhuang	H6	Rwd, Mjt	Chen and Xie (1993a, b)
Catherwood	L6	Rwd, Mjt	Coleman (1977)
Coorara	L6	Rwd, Mjt	Smith and Mason (1970)
Coolaman	L6	Rwd, Mjt	Smith and Mason (1970)
Roy	L5	Rwd, Mjt	Xie et al. (2006)
GRV 052174	L6	Rwd, Lgt	Feng (2011)
GRV 052246	L5/6	Rwd, Lgt	Feng (2011)
GRV 053492	L4	Rwd, Grt	Feng (2011)
S-98222	L6	Wds, Jdt	Ozawa et al. (2009)
GRV-05244	L5/6	Rwd	Feng (2011)
NWA-757	LL6	Rwd	Bishoff (2002)
Y-791099	LL5/6	Jdt	Kimura et al. (2001)
Y-791108	LL5/6	Jdt	Kimura et al. (2001)
Y-791141	LL5/6	Jdt	Kimura et al. (2001)

[a]*Rgt* ringwoodite, *Mjt* majorite, *Lgt* lingunite, *Akm* akimotoite, *Pvt* perovskite, *Mwt* magnesiowüstite, *grt* garnet, *Tu* tuite, *Xie* xieite, *CF* CaFe$_2$O$_4$-polymorph of chromite, *A-phase* high-pressure phase of chlorapatite, *Wds* wadsleyite, *Ca-Mjt* Ca-rich majorite, *Fe-Sp* Fe$_2$SiO$_4$-spinel, *Stv* stishovite, *Jdt* jadoite
[b]Also include Chen et al. (2003b, 2004, 2008)

oxide mineral ilmenite, which might have been dissolved in the shock melt. These findings are the unique for all studied chondrites, because the shock melt veins in many chondrites contain not only high-pressure mineral phases but also fragments of intact olivine, pyroxene, plagioclase and chromite.

4.3.4 Three New High-Pressure Minerals in Veins

Among the ten high-pressure mineral phases found in Suizhou shock veins, seven were first identified in other L6 chondrites. These are ringwoodite (the spinel-structured polymorph of olivine), majorite (the garnet-structured polymorph of pyroxene), akimotoite (the ilmenite-structured polymorph of pyroxene), vitrified perovskite (the perovskite-structured polymorph of pyroxene), lingunite (the hollandite-structured polymorph of plagioclase), majorite-pyrope garnet in solid solution, and magnesiowüstite. The other three new high-pressure phases we found in the Suizhou veins are tuite, the high-pressure polymorph of whitlockite (Xie et al. 2002, 2003), xieite, the $CaTi_2O_4$-structured polymorph of chromite (Chen et al. 2003a, b, 2008), and the CF phase, the $CaFe_2O_4$-structured polymorph of chromite (Chen et al. 2003b). Tuite is considered as the first fully identified high-pressure phosphate mineral ever found in natural materials. Xieite and the CF-phase are two post-spinel high-pressure phases that were predicted 40 years ago by Reid and Ringwood (1969, 1970) who proposed orthorhombic $CaFe_2O_4$-type and $CaTi_2O_4$-type structures as the top candidates for "postspinel" transitions in the Earth's mantle. So the finding of these phases in the Suizhou meteorite is of significance also for the study of Earth's mantle mineralogy.

4.4 Comparison with Melt Veins in NWA 3171 Chondrite

Besides the very thin and parallel melt veins that we observed in the Suizhou chondrite, Erin Walton observed melt veins in the NWA 3171 chondrite that are also thin, and cut across the entire stone as roughly straight parallel fractures (Fig. 4.8, personal communication). However, The interior of these veins in the NWA 3171 chondrite have a different texture and mineralogy. These veins are glass dominated, and no high-pressure minerals were found in them, so it might be better to say that the thin, straight and parallel nature of the veins in the Suizhou chondrite is distinct from most others and that the mineralogy of the veins is unique.

Fig. 4.8 Photograph of the NWA 3171 chondrite with thin, roughly parallel veins transecting the entire sample (From Erin Walton, personal communication)

4.5 Summary

1. The shock-produced melt veins in the Suizhou meteorite morphologically are the simplest, straightest and thinnest ones among all known shock-vein-bearing meteorites.
2. The narrow, fracture-like features in the Suizhou meteorite are shock-induced solid melt veins of chondritic composition and consist of fully crystalline materials without glassy material.
3. The Suizhou shock veins contain more abundant high-pressure mineral species than all other known vein-bearing chondrites. All rock-forming and almost all accessory oxide and phosphate minerals in the Suizhou shock veins were transformed to their high-pressure polymorphs, and no fragments of the precursor minerals were observed in veins. Among the ten high-pressure mineral phases identified in the Suizhou veins, three are new high-pressure minerals, namely, tuite, xieite and the CF-phase.

References

Ashworth JR (1985) Transmission electron microscopy of L-group chondrites, 1, Natural shock effects. Earth Planet Sci Lett 73:17–32

Binns RA, Davis RJ, Read SJB (1969) Ringwoodite, a natural (Mg, Fe)2SiO4 spinel in the Tenham meteorite. Nature 221:943–944

Bishoff A (2002) Discovery of purple-blue ringwoodite within shock veins of an LL6 ordinary chondrites from Northwest Africa. Lunar Planet Sci 33:1264

Boctor NZ, Bell PM, Mao HK (1982) Petrology and shock metamorphism of Pampa del Infierno chondrite. Geochim Cosmochim Acta 46:1903–1911

Chen M, Xie XD (1993a) The shock effects of olivine in the Yanzhuang chondrite. Acta Mineralogica Sinica 13(2):109–114

Chen M, Xie XD (1993b) The shock effects of orthopyroxene in the heavily shocked meteorites. Chin Sci Bull 38(12):1025–1027

Chen M, Wopenka B, Xie XD et al (1995) A new high-pressure polymorph of chlorapatite in the shocked chondrite Sixiangkou(L6). Lunar Planet Sci 26:237–238

Chen M, Sharp TG, El Goresy A et al (1996a) The majorite—pyrope + magnesiowustite assemblage: constrains on the history of shock veins in chondrites. Science 271:1570–1573

Chen M, Wopenka B, El Goresy A (1996b) High-pressure assemblage in shock melt vein in Peace River (L6) chondrite: compositions and pressure-temperature history. Meteoritics, 31(Suppl.):A27

Chen M, Xie XD, El Goresy A et al (1998a) Cooling rates in the shock veins of chondrites: constraints on the (Mg, Fe)$_2$SiO$_4$ polymorph transformations. Sci China Series D 41:522–552

Chen M, Xie XD, El Goresy A (1998b) Olivine plus pyroxene assemblages in the shock veins of the Yanzhuang chondrite: constraints on the history of H-chondrites. Neus Jahrbuch für Mineralogie 3:97–110

Chen M, Shu JF, Xie XD et al (2003a) Natural CaTi$_2$O$_4$-structured FeCr$_2$O$_4$ polymorph in the Suizhou meteorite and its significance in mantle mineralogy. Geochim Cosmochim Acta 67 (20):3937–3942

Chen M, Shu J, Mao HK et al (2003b) Natural occurrence and synthesis of two new postspinel polymorphs of chromite. Proc Natl Acad Sci USA 100(25):14651–14654

Chen M, El Goresy A, Frost D et al (2004) Melting experiments of a chondritic meteorite between 16 and 25 Gpa: Implications for Na/K fractionation in a primitive chondritic Earth's mantle. Eur J Miner 16:201–211

Chen M, Xie XD (2008) Two distinct assemblages of high-pressure liquidus phases in shock veins of the Sixiangkou meteorite. Meteor Planet Sci 2008(43):823–828

Chen M, Shu JF, Mao HK (2008) Xieite, a new mineral of high-pressure FeCr$_2$O$_4$ polymorph. Chin Sci Bull 53:3341–3345

Coleman LC (1977) Ringwoodite and majorite in the Catherwood meteorite. Canad Mineral 15:97–101

Dodd RT, Jarosewich E (1979) Incipient melting and shock classification of L-group chondrites. Earth Planet Sci Lett 44:335–340

Dodd RT, Jarosewich E, Hill B (1982) Petrogenesis of complex veins in the Chantonnay (L6f) chondrite. Earth Planet Sci Lett 59:364–374

Feng L (2011) The research of shock metamorphism and high-pressure minerals in Antarctic ordinary chondrites. Ph.D. Dissertation, Institute of Geology and Geophysics, Chinese Academy of Science, pp 113

Gillet P, Chen M, Dubrovinsky L et al (2000) Natural NaAlSi$_3$O$_8$ –hollandite in the Sixiangkou Meteorite. Science 287:1633–1637

Kimura M, El Goresy A, Suzuki A et al (1999) Heavily shocked Antarctic H-chondrites: petrology and shock history. Antarctic Meteorites 24:67–68

Kimura M, Suzuki A, Kondo T et al (2000) The first discovery of high-pressure polimorphs, jadeite, hollandite, wadsleyite and majorite, from an H-chondrite, Y-75100. Antarctic Meteorites 25:41–42

Kimura M, Suzuki A, Ohtani E et al (2001) Raman petrology of high-pressure minerals in H, L, LL, and E-chondrites. Meteor Planet Sci 36(Suppl):A99

Kimura M, Chen M, Yoshida Y et al (2003) Back-transformation of high pressures in a shock melt vein of an H-chondrite during atmospheric passage: Implications for the survival of high-pressure phases after decompression. Earth Planet Sci Lett 217:141–150

Madon M, Poirier JP (1979) Dislocation in spinel and garnet high-pressure polymorphs of olivine and pyroxene: implications for Mantle rheology. Science 207:66–68

Ohtani E, Kimura Y, Kimura M et al (2006) High-pressure minerals in shocked L6-chnodites: constraints on impact conditions. Shock Waves 16:45–52

Ozawa S, Ohtani E, Miyahara M et al (2009) Transformation textures, mechanisms of formation of high-pressure mineral in shock melt veins of L6 chondrites, and pressure-temperature conditions of the shock events. Meteor Planet Sci 44:1771–1786

Price GD, Putnis A, Agrell SO (1979) Electron petrography of shock-produced veins in the Tenham chondrite. Contrib Mineral Petrol 71:211–218

Reid AF, Ringwood AE (1969) Newly observed high pressure transformation in Mn_3O_4, $CaAl_2O_4$, and $ZrSiO_4$. Earth Planet Sci Lett 6:205–208

Reid AF, Ringwood AE (1970) The crystal chemistry of dense M_3O_4 polymorph: high pressure Ca_2GeO_4 of K_2NiF_4 structure type. J Solid State Chem 1:557–565

Rubin AE (1985) Impact melt products of chondritic material. Rev Geophys 23:277–300

Semenenko VD, Golovka NV (1994) Shock-induced black veins and organic compounds in ordinary chondrites. Geochim Cosmochim Acta 58:1525–1535

Sharp TG, Chen M, El Goresy A (1996) Microstructures of high-pressure minerals in shocked chondrites: Constraintson the duration of shock events. Meteor Planet Sci 31:A127

Sharp TG, Chen M, El Goresy A (1997a) Mineralogy and microstructures of shock-induced melt veins in the Tenham (L6) chondrite. Lunar Planet Sci 28:1283–1284

Sharp TG, Lingemann CM, Dupas C et al (1997b) Natural occurrence of $MgSiO_3$-ilmenite and evidence for $MgSiO_3$–perovskite in a shocked L chondrite. Science 277:255–352

Smith JV, Mason B (1970) Pyroxene-garnet transformation in Coorara meteorite. Science 168:822–823

Steel IM, Smith JV (1978) Coorara and Coolamon meteorites: ringwoodite and mineralogical differences. Lunar Planet Sci IX:1101–1103

Stöffler D, Keil K, Scott ED (1991) Shock metamorphism of ordinary chondrites. Geochim Cosmochim Acta 55:3854–3867

Tomioka N, Fujino K (1997) Natural (Mg, Fe) SiO3-ilmenite and—perovskite in the Tenham meteorite. Science 277:1084–1086

Xie ZD, Sharp TG (2007) Host rock solid-state transformation in a shock-induced melt vein of Tenham L6 chondrite. Earth Planet Sci Lett 254:433–445

Xie XD, Li ZH, Wang DD et al (1991) The new meteorite fall of Yanzhuang—A severely shocked H6 chondrite with black molten materials (Abstract). Meteoritica 26:411

Xie XD, Chen M, Wang DQ (2001a) Shock-related mineralogical features and P-T history of the Suizhou L6 chondrite. Eur J Miner 13(6):1177–1190

Xie XD, Chen M, Dai CD et al (2001b) A comparative study of naturally and experimentally shocked chondrites. Earth Planet Sci Lett 187:345–356

Xie XD, Chen M, Wang DQ et al (2001c) $NaAlSi_3O_8$-hollandite and other high-pressure minerals in the shock melt veins of the Suizhou L6 chondrite. Chin Sci Bull 46(13):1121–1126

Xie ZD, Tomioka N, Sharp TG (2002) Natural occurrence of Fe_2SiO_4-spinel in the shocked Umbarger L6 chondrite. Am Mineral 87:1257–1260

Xie XD, Minitti ME, Chen M et al (2003) Tuite, γ-$Ca_3(PO4)2$, a new phosphate mineral from the Suizhou L6 chondrite. Eur J Min 15:1001–1005

Xie XD, Chen M, Wang DQ (2005a) Two types of silicate melts in naturally shocked meteorites. Papers and abstracts of the 5th annual meeting of IPACES, Guangzhou, 12–14

Xie XD, Chen M, Wang DQ et al (2006) Melting and vitrification of plagioclase under dynamic high pressures. Acta Petrologica Sinica 22:503–509 (In Chinese with an English abstract)

Xie XD, Sun ZY, Chen M (2011) The distinct morphological and petrological features of shock melt veins in the Suizhou L6 chondrite. Meteor Planet Sci 46:459–469

Chapter 5
Mineralogy of Suizhou Shock Veins

5.1 General Remarks

The shock metamorphic effects observed in meteorites can be described in terms of either deformation or transformation or some combination of the two. As mentioned in Chap. 3, the deformation effects include fracturing, plastic deformation, twinning, and mosaicism within constituent minerals. Planar deformation features (PDFs) have been attributed to deformational processes, but they have also been shown to contain transformed material, either diaplectic glass or high-pressure phases (Xie and Chao 1987; Goltrant et al. 1991; Bowden 2002).

Transformational effects observed in shocked meteorites include shock melting, which commonly result in localized melt veins and pockets, transformation of minerals to high-pressure polymorphs, formation of diaplectic glass, and crystallization of highly deformed material. The high-pressure minerals that occur in shocked meteorites formed by either crystallization of silicate liquids in shock melt veins and pockets (Chen et al. 1996a; Xie and Chen 2009) or solid-state transformation of the constituent minerals in meteorites (Chen et al. 1996a). Solid-state phase transformation can provide important constraints on shock conditions of a meteorite, but transformation pressures are difficult to calibrate accurately because of kinetic effects and the heterogeneous nature of the initial transient shock pressure (Sharp and DeCarli 2006). Crystallization of chondritic melt provides an alternative means of constraining the crystallization pressure, which can be related to the shock pressure of the sample.

High-pressure minerals are common in and around shock melt veins in highly shocked meteorites (Chen et al. 1996a; Xie et al. 2001a). Ringwoodite, the spinel-structured polymorph of olivine, was firstly discovered in the Tenham L6 chondrite (Binns et al. 1969). However, this new high-pressure mineral had already been observed in the Coorara chondrite one year before by Mason et al. (1968), but

© Springer-Verlag Berlin Heidelberg and Guangdong Science
& Technology Press Co., Ltd. 2016
X. Xie and M. Chen, *Suizhou Meteorite: Mineralogy and Shock Metamorphism*,
Springer Geochemistry/Mineralogy, DOI 10.1007/978-3-662-48479-1_5

it was misidentified as garnet based on X-ray diffraction data from the majorite garnet in the sample. Smith and Mason (1970) clarified the mistake when they published the discovery of majorite garnet in Coorara. The majorite that they describe occurred in a fine-grained mixture of garnet, Fe oxide, and iron, which must have been crystallized from chondritic melt such as that described by Chen et al. (1996a). The fact that these garnets had higher concentrations of Na, Al, and Cr than the orthopyroxenes in the sample confirms that they crystallized from the melt. In the same paper, Smith and Mason described an isotropic phase with an orthopyroxene composition that they speculated was also majorite. It was clear from this early work that ringwoodite and majorite were observed in close association with melt veins in shocked chondrites.

Putnis and Price (1975) used transmission electron microscopy to characterize the microstructures of ringwoodite in Tenham and subsequently discovered the β-spinel polymorph of olivine, which they later named wadsleyite. In 1996, that is 25 years later since the time of ringwoodite and majorite discoveries, Chen et al. (1996a) found two distinguished high-pressure assemblages in shock melt veins in the Sixiangkou L6 chondrite: (i) majorite–pyrope solid solution plus magnesiowüstite that crystallized at high pressures and temperatures from a shock-induced silicate melt of bulk Sixiangkou composition and (ii) ringwoodite plus low-calcium majorite that were produced by solid-state transformation of olivine and low-calcium pyroxene. After that, numerous other high-pressure minerals, such as hollandite-structured plagioclase, $MgSiO_3$ perovskite glass, akimotoite (an ilmenite-structured polymorph of $MgSiO_3$), tuite (a high-pressure polymorph of whitlockite), and two post-spinel polymorphs of chromite, have been discovered in shocked chondrites. It should be pointed out that in nearly all cases, the high-pressure polymorphs only occur within shock melt veins or adjacent to them.

The Suizhou chondrite contains a few thin melt veins of 0.02–0.20 mm in width (Fig. 5.1). According to previous investigators, FeNi metal and troilite in shock veins occur in the form of eutectic blebs (Shen and Zhuang 1990), and the silicate minerals in the veins were only brecciated by shock, and no high-pressure minerals were found in the veins (Wang 1993). On the other hand, most high-pressure-phase-bearing L6 chondrites studied so far contain thick shock veins up to one or several millimeters in width, such as those in Sixiangkou, Peace River (Chen et al. 1996b, 1998), and Tenham meteorites (Sharp et al. 1997a). Two assemblages of high-pressure phases, coarse-grained ringwoodite (or wadsleyite) + majorite + lingunite (hollandite-structured plagioclase), and fine-grained majorite–pyrope$_{ss}$ + magnesiowüstite were found in these thick shock melt veins. However, high-pressure phases are not reported from those L6 chondrites that shock veins are very thin and very poorly developed. So it is reasonable to assume that the very thin shock-vein-bearing Suizhou L6 chondrite would not have contained high-pressure phases.

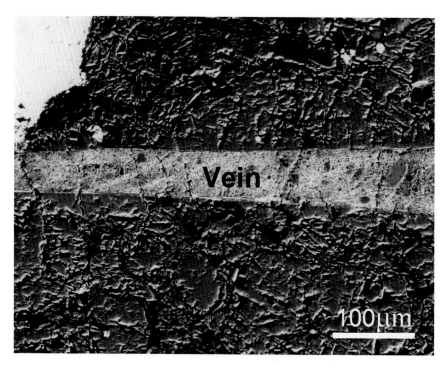

Fig. 5.1 BSE image showing a solid shock melt vein of 70–90 μm in width in the Suizhou meteorite. Note the sharpness of the vein boundaries with its surrounding unmelted chondritic rock and the presence of some coarse-grained high-pressure minerals distributed in the fine-grained vein matrix

However, as we have described in Chap. 4, our recent micro-mineralogical studies did reveal that the shock melt veins of the Suizhou chondrite are really full of high-pressure phases. Furthermore, similar to the thick shock melt veins in many other L6 chondrites (Chen et al. 1996a; Sharp et al. 1997a), two types of high-pressure mineral assemblages were also developed in the very thin Suizhou veins (Figs. 5.1 and 5.2), namely the coarse-grained assemblage of ringwoodite + majorite + akimotoite + perovskite + lingunite ($NaAlSi_3O_8$ hollandite) + tuite (the high-pressure polymorph of whitlockite) + xieite and the CF phase (two high-pressure polymorphs of chromite), and the fine-grained assemblage of matrix minerals majorite–pyrope solid solution + magnesiowüstite + microcrystalline ringwoodite (Xie et al. 2001a, 2002, 2003, 2011a, b; Chen et al. 2003a, b). These two types of high-pressure mineral constitute up to 90 % of materials in veins by volume. The other 10 % constituents of veins are fine-grained metal–troilite eutectic intergrowths and blebs which fill the interstices of majorite–pyrope$_{ss}$ or in the form of metal–sulfide veinlets (Xie et al. 2001a). This means that almost all minerals in the Suizhou shock veins have transformed to their high-pressure polymorphs. This is really a very rare and unique case that up to 10 minerals in the Suizhou shock

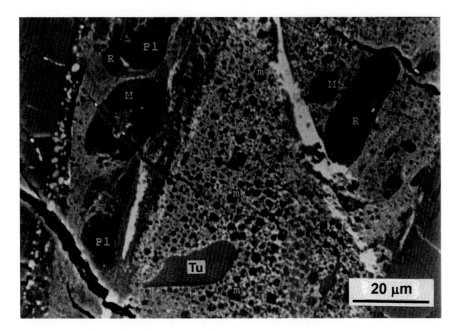

Fig. 5.2 BSE image of a shock melt vein showing the coarse-grained high-pressure assemblage consisting of ringwoodite (*R*), majorite (*M*), NaAlSi$_3$O$_8$ hollandite (*Pl*), and tuite (*Tu*) and the fine-grained matrix consisting of majorite–pyrope garnet (*m*) and FeNi metal + FeS in intergrowth (*white colored*)

melt veins have transformed to their high-pressure polymorphs. The only one mineral that we could not find its high-pressure polymorph is ilmenite. In this chapter, we describe the characteristics of the high-pressure mineral phases in Suizhou melt veins.

5.2 Coarse-Grained High-Pressure Mineral Phases

5.2.1 Ringwoodite, the Spinel-Structured (Mg,Fe)$_2$SiO$_4$

The γ-phase of olivine was firstly obtained at high-pressure and high-temperature experiments by Ringwood and Major (1966). Prewitt (1998) pointed out that "the only known natural occurrence of a (Mg,Fe)$_2$SiO$_4$ spinel is in the shocked meteorites". The high-pressure spinel polymorph of (Mg,Fe)$_2$SiO$_4$ olivine was first reported by Binns et al. (1969) and Binns (1970) in black shock veins in the Tenham L6 ordinary chondrite and was named as "ringwoodite" after the name of the famous Australian petrologist, Professor Ringwood. Ringwoodite has now been widely recognized in shock melt veins of many other highly shocked ordinary chondrites of shock stage 6, such as Catherwood (Coleman 1977), Sixiangkou

(Chen et al. 1996a), Peace River (Chen et al. 1996b), Mbale (Chen et al. 1998), Suizhou (Xie et al. 2001a), and some other meteorites, as well as in the shock melt veins of a few H-group chondrites, such as Yanzhuang (Chen and Xie 1993) and an Antarctic meteorite Y-75100 (Kimura et al. 2000).

The formula for the spinel-type structure is B_2AO_4. The crystal structure of ringwoodite can be viewed as the diagonal crisscrossing chains of MgO octahedra linked together by isolated SiO_4 tetrahedra. The oxygen anions are in a cubic closely packed arrangement with the B cation in octahedral sites and the A cation in tetrahedral. The octahedra form edge-sharing chains that are linked together by the isolated tetrahedra. All atoms are in special positions, and the only variable for the structure is the cubic cell edge and the x position of the oxygen atom (Prewitt and Downs 1998). As far as the Earth's interior is concerned, $(Mg,Fe)_2SiO_4$ ringwoodite is the most important mineral phase in the mantle's transition zone. Under the condition of lower mantle, ringwoodite transforms to $(Mg,Fe)O$ magnesiowüstite + $(Mg,Fe)SiO_3$ perovskite.

During our recent micro-mineralogical studies on the shock melt veins of the Suizhou L6 chondrite, we also found quite a lot of ringwoodite grains in the shock melt veins of this meteorite (Xie et al. 2001a). The physical and chemical properties of the Suizhou ringwoodite are described in the following sections.

Occurrence

The Suizhou polycrystalline grains of ringwoodite of 10–50 μm in length are smooth, rounded, and unfractured. SEM and LRM investigations showed that these grains are single-phase ringwoodite (Figs. 5.2, 5.3 and 5.4). In comparison with the heavily deformed and fractured grains of host olivine in the Suizhou unmelted chondritic rock, the smooth and unfractured features of the large ringwoodite grains in Suizhou shock veins imply that these grains experienced strong shock compression, and the rounded outline of the ringwoodite fragments is indicative of partial resorption into the fine-grained molten matrix during the shock event.

Sometimes, we observe a few fractures or cracks penetrating the ringwoodite grains in the Suizhou shock veins (Figs. 5.2, 5.3, and 5.4), but they are not produced by shock compression but by cooling shrinkage after thermal expansion of the melt veins.

Chemical Composition

Electron microprobe analyses revealed that the Suizhou ringwoodite grains have the same chemical composition as the olivine outside the melt veins (Table 5.1). It is clear that these ringwoodite grains must have formed directly from olivine through isochemical solid-state phase transformation under high pressures and no additional elements were incorporated with these grains from the surrounding matrix melt.

Raman Spectroscopy

A micro-Raman spectrum obtained from a 30-μm-long grain of the Suizhou ringwoodite displays two strong Raman peaks at 796 and 841 cm^{-1} (Fig. 5.5a). This spectrum is comparable with that of γ-Mg_2SiO_4 (spinel) (McMillan and Akaogi 1987) and is different with that of olivine outside the veins, whose spectrum shows

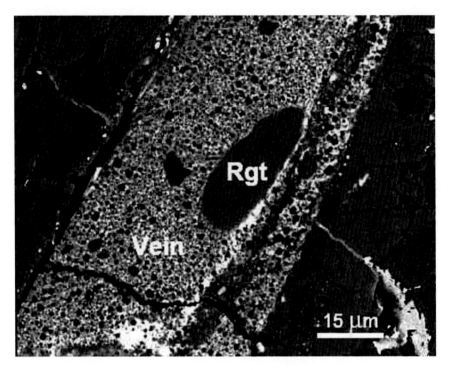

Fig. 5.3 BSE image showing an elongated ringwoodite grain (*Rgt*) in a shock vein of the Suizhou meteorite. Note the fine-grained majorite–pyrope garnet (*dark gray*) and the metal–sulfide intergrowth (*white*) in the vein

two main peaks at 821 and 851 cm^{-1}. Although the two respective Raman peaks for both olivine and ringwoodite are assigned to the symmetric and asymmetric stretching vibrations of SiO_4 tetrahedra, ringwoodite, having higher symmetry and less distortion of SiO_4 tetrahedra than olivine, shows the decreasing of Raman peak height and shifting of Raman peaks to the direction of smaller wave numbers.

X-ray Diffraction

The X-ray diffraction pattern of the Suizhou ringwoodite was obtained by using the in situ micro-diffraction analysis. Since the ringwoodite grain is surrounded by other fine-grained minerals of the vein matrix, the result shows the diffraction lines of ringwoodite together with lines of majorite–pyrope garnet, kamacite, and troilite. Among them, the following lines belong to ringwoodite (in Å): 2.446 (100), 1.431 (50), 2.026 (35), 2.878 (20), 1.561 (20), 1.059 (20), and 1.056 (10). These data are consistent with those of standard ringwoodite of JCPDS No. 21-1258.

Olivine–Ringwoodite Transformation Mechanism

It was revealed that the ringwoodite in very highly shocked chondrites occurs as polycrystalline aggregates that have the same chemical composition as the olivine in the same samples (Chen et al. 1996a; Langenhorst et al. 1995; Xie et al. 2001a). This implies that ringwoodite formed from olivine via a solid-state transformation

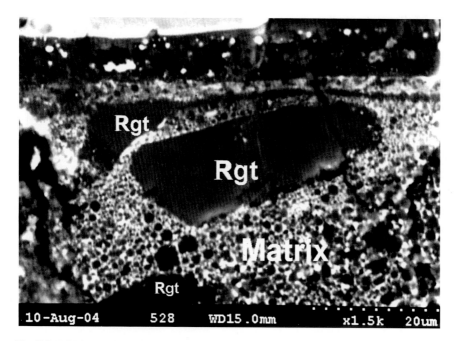

Fig. 5.4 BSE image showing some large polycrystalline ringwoodite grains (*Rgt*) surrounded by fine-grained matrix (*Matrix*) in a shock vein of the Suizhou meteorite

Table 5.1 Chemical composition of ringwoodite in the Suizhou shock veins (wt%)

	SZ-Rwt-1	SZ-Rwt-2	SZ-Rwt-3	SZ-Rwt	Ringwoodite average	Olivine average
SiO_2	38.202	38.439	38.211	38.706	38.389	38.249
TiO_2	0.003	0.000	0.056	0.000	0.015	0.015
Al_2O_3	0.007	0.000	0.008	0.002	0.004	0.000
Cr_2O_3	0.045	0.021	0.082	0.027	0.044	0.013
MgO	38.542	38.771	38.471	38.964	38.687	39.174
CaO	0.028	0.052	0.046	0.053	0.045	0.006
MnO	0.488	0.542	0.494	0.397	0.480	0.481
FeO	23.356	23.532	22.973	23.171	23.258	22.343
NiO	0.000	0.063	0.038	0.060	0.040	0.039
Na_2O	0.000	0.000	0.000	0.000	0.000	0.014
K_2O	0.015	0.000	0.014	0.016	0.011	0.011
Total	100.686	101.420	100.393	101.396	100.973	100.345
Fa	25.25	25.27	24.98	24.93	25.11	24.12
Fo	74.26	74.20	74.54	74.71	74.43	75.3
Mo	0.04	0.06	0.06	0.07	0.06	0.01
Li	0.00	0.07	0.04	0.06	0.04	0.04
Te	0.53	0.59	0.54	0.43	0.52	0.52

All values were determined by EPMA in wt%
Fa—fayalite, Fo—forsterite, Mo—monticellite, Li—liebenbergite, Te—tephroite

Fig. 5.5 Raman spectra of
ringwoodite in Suizhou shock
melt veins (*a*), γ-Mg$_2$SiO$_4$
(spinel) (*b*), and olivine
outside the Suizhou veins (*c*)

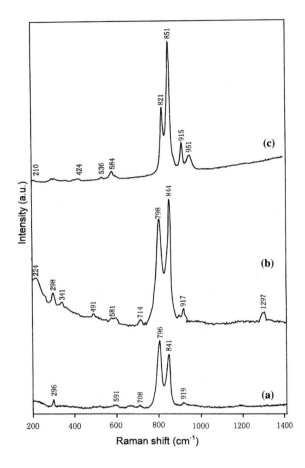

mechanism during shock compression (Chen et al. 1996a). TEM examination of the
polycrystalline ringwoodite in chondrites generally shows randomly oriented
ringwoodite crystallites that range from about one hundred nanometers (Price et al.
1979) to several micrometers (Chen et al. 1996a). The random orientation and
homogeneous distribution of ringwoodite crystallites indicate homogeneous
intracrystalline nucleation throughout the olivine rather than heterogeneous nucleation on grain boundaries, which is dominate mechanism at pressures closer to the
equilibrium phase boundary (Kerschofer et al. 2000; Mosenfelder et al. 2001). The
presence of small amounts of glassy material in ringwoodites from Tenham
meteorite (Price et al. 1979) has been interpreted as remnants of a prograde
high-density olivine glass that was an intermediate phase in the transformation of
olivine to ringwoodite. However, such glassy phases have not been reported in
more recent studies, and static high-pressure experiments never demonstrated that
the olivine–ringwoodite phase transition path goes via an intermediate glass phase
(Agee et al. 1995; Zhang and Herzberg 1994). In most samples of highly shocked
chondrites, the ringwoodite composition is constant, implying that there was no

Fe–Mg exchange during the transformation, and therefore, the ringwoodite crystallites grew by interface growth rather than diffusion-controlled growth (Sharp and DeCarli 2006).

Chen et al. (2004b) reported the first natural occurrence of ringwoodite lamellae in the olivine grains inside and in the areas to the shock veins of the Sixiangkou meteorite. They found that inside the veins where pressure and temperature were higher than elsewhere, ringwoodite lamellae formed parallel to the {101} planes of olivine, whereas outside they lie parallel to the (100) plane of olivine. The lamellae replaced the host olivine from a few percent to complete. They assumed that formation of these lamellae is related to a diffusion-controlled growth of ringwoodite along shear-induced planar defects in olivine. Hence, these ringwoodite lamellae show distinct growth mechanism.

It has been revealed that the ringwoodite grains in the shock melt veins of the Suizhou chondrite have the same chemical composition as the olivine in the unmelted chondritic rock (Table 5.1). We consider that these ringwoodite grains must have formed directly from olivine through solid-state phase transformation under high pressures. The polycrystalline nature of the Suizhou ringwoodite grains and the constant composition for different ringwoodite grains indicate the homogeneous features in chemical composition of ringwoodite crystallites within grains. Therefore, it is reasonable to assume that olivine–ringwoodite transformation mechanism in the Suizhou meteorite can be interpreted to be the homogeneous intracrystalline nucleation of ringwoodite crystallites throughout the host olivine grain and followed by the interface-controlled growth of crystallites in the grain.

P-T Conditions

Ringwoodite is the most abundant mineral in the Earth's transition zone between the 520 and 660 discontinuities (Anderson and Bass 1986). Static high-pressure kinetic experiments have shown that dry hot-pressed San Carlos olivine transforms to ringwoodite, on an observable timescale, only above 900 °C at 18–20 GPa (Kerschofer et al. 1996, 1998, 2000). According to the conditions of formation of other high-pressure minerals of the veins, such as majorite, hollandite, and magnesiowüstite, it was assumed that ringwoodite in the Suizhou veins formed at 18–23 GPa and 1800–1900 °C. This pressure regime corresponds to the pressure range of the mantle transition zone, e.g., 400–660 km deep in the Earth's mantle.

5.2.2   Majorite, the Garnet-Structured (Mg,Fe)SiO$_3$

Majorite is the high-pressure polymorph of pyroxene with garnet structure. This polymorph (MgSiO$_3$ garnet) is a stable phase in the pressure range 19–24 GPa at the temperature between 1700 and 2600 °C (Kato and Kumazawa 1985). Smith and Mason (1970) reported the first natural occurrence of majorite in a veinlet in the Coorara meteorite, produced by high-pressure transformation of low-Ca pyroxene to the higher density garnet structure and was named after the Professor A. Major. The majorite is extremely fine-grained and difficult to analyze, but has an Mg/ (Mg + Fe) ratio of ~0.25. Coleman (1997) also described the occurrence of majorite in Catherwood (L6) chondrite and reported compositional data. Price et al. (1979) examined the microstructures of majorite in shock veins in the Tenham L6 chondrite. Since then, the low-Ca majorite has been discovered in the shock melt veins of a series of L-group chondrites, such as Sixiangkou (Chen et al. 1996a), Peace River (Chen et al. 1996b), Mbale (Chen et al. 1998b), Suizhou (Xie et al. 2001a), and some other L6 chondrites, as well as in the shock melt veins of a few H-group chondrites, such as Yanzhuang (Chen and Xie 1993a) and an Antarctic meteorite Y-75100 (Kimura et al. 2000). Low-Ca majorite in shock veins of the Sixiangkou L6 chondrite occurs as rounded polycrystalline grains, 15–300 μm in size, associated with ringwoodite and lingunite (the hollandite-structured plagioclase). This majorite contains a subgrain dislocation microstructure consistent with dislocation climb. The majorite has a composition essentially identical to unshocked low-Ca pyroxene in the Sixiangkou unmelted chondritic rock and probably formed by solid-state transformation of low-Ca pyroxene.

The majorite has space group $Ia3d$ and occurs as equant grains. The equant grains are 50–500 nm in size and have 120° grain junctions. The crystal structure of majorite can be viewed as a rather rigid framework of corner-sharing octahedra (B-site) and tetrahedra, with the A-site atom located in interstices that have a dodecahedral shape.

During our recent micro-mineralogical studies on the shock melt veins of the Suizhou L6 chondrite, we also found quite a lot of majorite grains in the shock melt veins of this meteorite (Xie et al. 2001a). The brief introductions of the Suizhou majorite are described in the following sections.

Occurrence

Polycrystalline grains of low-Ca majorite occur in the shock melt veins of the Suizhou meteorite with the grain sizes ranging from 8 to 50 μm in diameter (Figs. 5.2 and 5.6). These grains are also smooth and unfractured and have dark gray color under the SEM. Figure 4.6 is a BSE image of the Suizhou meteorite showing that the shock vein contains coarse-grained polycrystalline majorite, ringwoodite, and hollandite-structured polymorph of plagioclase, as well as the polycrystalline fine-grained matrix. It is understandable that this majorite is a pseudomorph after pyroxene and shows a fine-grained polygranular texture.

Chemical Composition

Electron microprobe analyses show that Suizhou low-Ca majorite has the same chemical composition as the low-Ca pyroxene outside the veins (Table 5.2). Hence, the coarse-grained low-Ca majorite is formed from low-Ca pyroxene through solid-state phase transition under high pressures. However, the electron microprobe analyses revealed that the idiomorphic fine-grained majorite garnets in Suizhou veins are richer in Al_2O_3, CaO, Na_2O, and Cr_2O_3 in comparison with that of low-Ca pyroxene outside the veins (Table 5.2). It is evident that these two kinds of majorite have different origins of their formation.

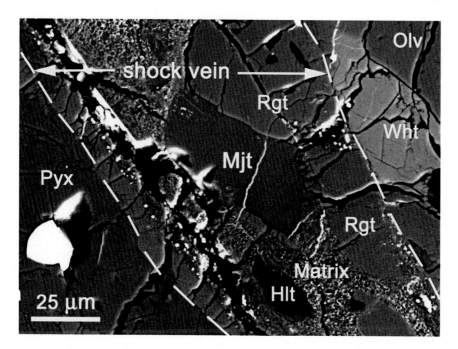

Fig. 5.6 A BSE image of the Suizhou meteorite showing that the shock melt vein contains coarse-grained, polycrystalline majorite (*Mjt*), ringwoodite (*Rgt*), and hollandite-structured polymorph of plagioclase (*Hlt*), as well as the polycrystalline fine-grained matrix (*Matrix*). Olv = olivine, Pyx = pyroxene, Wht = whitlockite (Chen et al. 2004c)

Table 5.2 Chemical composition of majorite in the Suizhou shock veins (wt%)

	SZ-Mjt-1	SZ-Mjt-2	SZ-Mjt-3	SZ-Mjt-4	Majorite average	Pyroxene average
SiO_2	55.965	56.458	55.068	55.695	55.797	55.775
TiO_2	0.158	0.192	0.164	0.252	0.191	0.162
Al_2O_3	0.153	0.187	0.156	0.148	0.123	0.158
Cr_2O_3	0.124	0.106	0.169	0.085	0.121	0.110
MgO	29.182	28.130	29.033	29.218	28.891	29.239
CaO	0.736	0.772	0.678	0.652	0.710	0.715
MnO	0.503	0.519	0.518	0.458	0.499	0.496
FeO	14.477	13.828	14.881	14.272	14.365	13.952
NiO	0.124	0.048	0.092	0.041	0.076	0.007
Na_2O	0.024	0.426	0.000	0.025	0.119	0.043
K_2O	0.007	0.063	0.018	0.003	0.023	0.009
Total	101.451	100.729	100.768	100.848	100.915	100.666
Fs	20.74	21.91	20.11	20.63	20.85	20.12
En	77.85	76.58	78.57	78.12	77.78	78.41
Wo	1.41	1.51	1.32	1.25	1.37	1.38

All values were determined by EPMA in wt%
Fs—ferrosite, En—enstatite, Wo—wollastonite

Fig. 5.7 Raman spectra of majorite (*a*), majorite–pyrope$_{ss}$ (*b*) in Suizhou melt veins, and pyroxene outside the Suizhou veins (*c*)

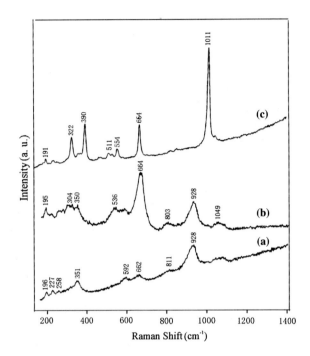

Raman Spectroscopy

The Raman spectrum of a Suizhou coarse-grained low-Ca majorite grain displays Raman peaks at 928, 662, 592, and 351 cm^{-1} (Fig. 5.7a). The Raman spectrum of another polycrystalline majorite grain in the Suizhou chondritic rock displays sharp strong bands at 926 and 590 cm^{-1} and weak bands at 315, 373, 654, 804, and 1056 cm^{-1}. The strong peak at 926–928 cm^{-1} corresponds to the stretching vibrations of the SiO$_4$ tetrahedra. Since the Raman spectra of majorite are significantly different for those of low-Ca pyroxene, we can easily identify the majorite phase from its precursor pyroxene by Raman spectroscopy.

X-ray Diffraction

The in situ micro-diffraction analysis on the Suizhou meteorite is shown in Fig. 5.8, in which tiny grains of majorite in a shock vein matrix are exposed to X-ray of 0.03-mm collimator with $\omega = 22°$ for 2 h. The result shows the diffraction lines of majorite with minor kamacite and troilite. The following lines belong to majorite (in Å): 2.574 (100), 2.878 (80), 1.541 (50), 2.455 (45), 1.598 (40), 2.261 (30), 2.350 (30), 2.037 (20), 1.868 (20), and 1.664 (15). These data are consistent with those of standard majorite of JCPDS No. 25-0483.

Low-Ca Pyroxene–Majorite Transformation Mechanism

It has been found that majorite in most shocked chondrites occurs as polycrystalline aggregates that have the same chemical composition as the low-Ca pyroxene in the unmelted chondritic rock (Chen et al. 1996a, Xie et al. 2001a). This implies that majorite formed from low-Ca pyroxene also via a solid-state transformation

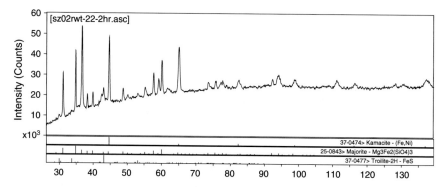

Fig. 5.8 X-ray micro-diffraction pattern of majorite in a shock vein matrix in the Suizhou meteorite (2θ-I profile showing the diffraction lines of majorite with minor kamacite and troilite)

mechanism during shock compression (Chen et al. 1996a). The mechanism of transformation of pyroxene to majorite appears to be the same as that of ring-woodite: homogeneous intracrystalline nucleation of majorite crystallites followed by interface-controlled growth (Sharp and DeCarli 2006). Price et al. (1979) suggested that shocked pyroxene first did not transform to majorite directly but transforms into a prograde glass from which majorite subsequently crystallized. However, such intermediate glassy phases have not been reported in more recently studies, and the static high-pressure experiments never demonstrated that the orthopyroxene–majorite phase transition path goes via an intermediate glass phase (Agee et al. 1995; Wang and Takahashi 2000; Chen et al. 2004a). Therefore, we firmly stand for a mechanism of homogeneous nucleation and interface-controlled growth in the solid state for the coarse-grained majorite in the shock melt veins of the Suizhou meteorite.

P-T Condition

High-pressure experiments indicated that $MgSiO_3$ majorite has a *P-T* stability field between 16 and 22.5 GPa and 1600–2500 °C (Gasparik 1992; Presnall and Gasparik 1990; Chen et al. 2004a), and the presence of majorite garnet plus ring-woodite constrains the *P-T* condition of 18–23 GPa and 1800–1900 °C (Agee et al. 1995). On the basis of coarse-grained high-pressure assemblage in the Suizhou shock melt veins, we believe that the Suizhou majorite formed at high pressure up to 23–24 GPa and temperature up to 1900–2000 °C. This pressure regime corresponds to the pressure range from the lower part of mantle transition zone to the upper part of the lower mantle, e.g., 600–700 km deep in the Earth's mantle.

5.2.3   Akimotoite, the Ilmenite-Structured (Mg,Fe)SiO₃

At high pressure and temperature, pyroxene transforms to high-pressure poly-morphs including majorite (the garnet-structured $(Mg,Fe)SiO_3$), akimotoite (the ilmenite-structured $(Mg,Fe)SiO_3$), and perovskite (the perovskite-structured (Mg, Fe)SiO_3. The ilmenite-structured $(Mg,Fe)SiO_3$, which was synthesized by Kawai et al. (1974) and identified as having an ilmenite structure by Liu (1976), is a potentially important mantle mineral that has not been previously found in terrestrial rocks. This high-pressure dense polymorph has only been found in the shock melt veins of chondrites, but not widely identified in these meteorites. Up to now, only a few meteorites were reported to contain very small amount of akimotoite.

Sharp et al. (1997a, b) examined shock melt veins in the Acfer 040 (L5-6, S6) chondrite using TEM technique and observed prismatic or plate-like grains with the composition $MgSiO_3$, with minor concentrations of FeO, Al_2O_3, Na_2O, and Cr_2O_3. Based on electron diffraction studies, the phase was unambiguously identified as ilmenite-structured $MgSiO_3$ with space group R-3, the first occurrence of this phase in nature. The $MgSiO_3$ ilmenite occurs with ringwoodite and an amorphous phase which may have been $MgSiO_3$ perovskite, but became amorphous during depressurization. Because $MgSiO_3$ ilmenite is not predicted as a stable phase at high pressure, Sharp et al. (1997a, b) argued that it crystallized metastably during post-shock decompression and heating.

Tomioka and Fujino (1997) also identified the mineral $MgSiO_3$ ilmenite in the shock-induced veins in the Tenham L6 chondrite using ATEM technique. This mineral is adjacent to clinopyroxene in fragments and formed aggregates with each grain <1.4 μm in length. The examined hexagonal structure, d-spacings, angles, space group (R-3), and systematic extinctions of electron diffraction patterns of this mineral are all consistent with those of ilmenite. Chemical analyses of $MgSiO_3$ ilmenite grains were carried out with an energy-dispersive analytical system attached to the ATEM and found that these grains have similar compositions to the adjacent clinopyroxene grains. $MgSiO_3$ ilmenite grains in the Tenham meteorite have two morphologies. One is granular-shaped (<0.4 μm in length), and the other is columnar (<1.4 μm in length). Both types of grains do not show any microstructures except a low density of dislocations. Tomioka and Fujino (1997) argued that the columnar-type ilmenite has the shear transformation mechanism for the clinopyroxene–ilmenite transition, because the topotaxial relation indicates that this process may have proceeded by the displacement of the closely packed layers of oxygen on (100) plane for clinopyroxene, whereas the granular ilmenite, which has no topotaxial relationship with clinopyroxene, would have formed by the nucleation and growth mechanism, probably under the slower cooling rates. This $MgSiO_3$-ilmenite was than approved as a new mineral with the mineral name akimotoite (Tomioka and Fujino 1999).

Ferroir et al. (2008) reported observations of four textural relationships between pyroxene and akimotoite in former pyroxene grains entrained in the shear melt vein and in pyroxene grains attached to the wall of the melt vein in the Tenham L6 chondrite. They suggested that akimotoite is mainly formed by solid-state transformation of former pyroxenes with subsequent diffusion of calcium, aluminum, and sodium from the chondritic melt of the shear melt vein.

The other occurrences of akimotoite in chondrites were reported in succession by Xie et al. (2005b) in the Suizhou chondrite and then by Zhang et al. (2006) and

Miyagima et al. (2007) in the Sixiangkou chondrite. Akimotoite was identified by its characteristic plate-like morphology in the TEM images, selected area electron diffraction patterns, and quantitative energy-dispersive X-ray spectra. The akimotoite-bearing fine-grained matrix of the melt vein is classified as a metal–troilite-poor lithology by Chen et al. (1996a). Akimotoite crystals range in size from 0.3 to 3 μm and the plate-like idiomorphic crystals are elongated parallel to the $(110)_{hex}$ direction. The coexisting phases are a majorite–pyrope solid solution, ringwoodite, FeNi metal, troilite (FeS), and a silicate glass. Akimotoite, which is considered to be a subsolidus phase in the Mg_2SiO_4–$MgSiO_3$ system (Gasparik 1992), crystallized with ringwoodite and majorite from the immiscible silicate and metal–sulfide liquids in the melt vein at very high temperatures and high pressures on the order of 2000 °C at 23 GPa. These values are estimated from the occurrence of the liquidus pair, majorite garnet and magnesiowüstite, at the central core of the Sixiangkou melt vein (Chen et al. 1996a). On the other hand, melting experiments in the CV3 chondrite Allende did not reveal akimotoite as a liquidus phase during cooling under equilibrium conditions (Agee et al. 1995). However, akimotoite is considered to have crystallized metastably at the margin of the melt vein where the cooling rate from high temperature was highest due to rapid heat transport to the matrix so that supercooling was large (Sharp et al. 1997a, b; Xie et al. 2006). The crystallization of akimotoite in the Sixiangkou chondrite is also inferred to have occurred as a result of supercooling of the melt at the margin of the melt vein, at a low oxygen fugacity (fO_2) as documented by the coexistence with FeNi metal.

During our recent micro-mineralogical studies on the shock melt veins of the Suizhou L6 chondrite, akimotoite was also identified in the shock-produced melt veins or in the area directly adjacent to the shock melt veins of this meteorite (Chen and Xie 2015). The physical and chemical properties of the Suizhou akimotoite are described in the following sections.

Occurrence

A most important occurrence of akimotoite in the Suizhou meteorite is to occur along the fractures inside pyroxene fragments enclosed in melt as irregular zones of akimotoite (Fig. 5.9). The width of akimotoite zones is about 2–4 μm. On the BSE image, akimotoite shows dark gray color which is little bit lighter than the very dark pyroxene. Another occurrence is to appear in the edge of pyroxene or fractures in these pyroxene grains in contact with the melt vein. This second occurrence of akimotoite in the Suizhou meteorite is in a granular texture where the crystal grains have a topotaxial relationship with the host pyroxene and form a rather narrow zone of randomly oriented akimotoite grains in between intact pyroxene and shock-produced $(Mg,Fe)SiO_3$ glass which is in direct contact with a shock melt vein (Fig. 5.10). The grain size of akimotoite is in the range of 2–4 μm. On the BSE image, it shows gray color which is also lighter than the pyroxene and the $(Mg,Fe)SiO_3$ glass.

Chemical Composition

Energy-dispersive spectroscopic analyses show that akimotoite in the Suizhou melt veins or in the area directly adjacent to Suizhou shock veins has an identical composition as its host low-Ca pyroxene (Table 5.3). The contents of the main

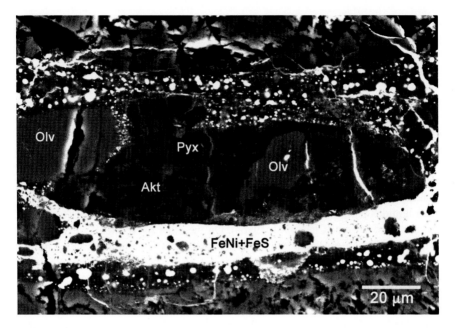

Fig. 5.9 BSE image of an area directly adjacent to a shock vein in Suizhou meteorite showing the occurrence of akimotoite (*Akt*) occurring along the fractures inside pyroxene (*Pyx*) fragments enclosed in melt. Olv = olivine, FeNi + FeS = metal + troilite intergrowth (Chen and Xie 2015)

constituents MgO, FeO, and SiO_2, as well as the minor constituents, such as MnO, CaO, Al_2O_3, Cr_2O_3, TiO_2, and K_2O, are almost identical for both akimotoite and the host pyroxene. It is clear that akimotoite in the Suizhou meteorite must have formed directly from pyroxene through isochemical solid-state phase transformation under high pressures.

Raman Spectroscopy

The Raman spectrum of the Suizhou akimotoite displays Raman bands at 798, 676, 614, 478, 406, and 344 cm^{-1} (Fig. 5.11a). The strong peak at 798 cm^{-1} corresponds to the stretching vibrations of the SiO_4 tetrahedra. Raman spectrum of akimotoite is distinct from its host orthopyroxene, which show clear Raman bands at 1014, 927, 793, 677, 658, 520, 407, 385, 335, and 233 cm^{-1} (Fig. 5.10b). Since the Raman spectrum of akimotoite is significantly different for that of low-Ca pyroxene, we can easily identify the akimotoite phase from its precursor pyroxene by Raman spectroscopy.

Low-Ca Pyroxene-Akimotoite Transformation Mechanism

In the transformation of pyroxene to akimotoite in the shock veins of the Tenham meteorite, akimotoite occurs in a granular texture as well as in columnar texture where the crystals have a topotaxial relationship with the pyroxene (Tomioka and Fujino 1997). The granular texture consists of randomly oriented crystallites from 100 to 200 nm, which is consistent with homogeneous intracrystalline nucleation

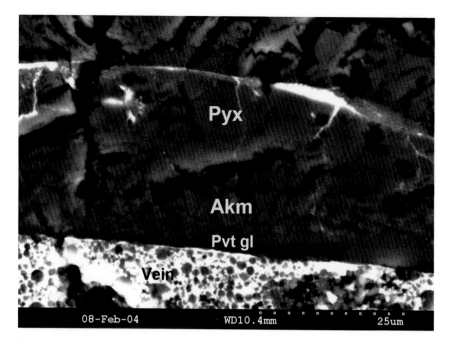

Fig. 5.10 BSE image showing the occurrence of granular akimotoite (*Akt*) in between the low-Ca pyroxene (*Pyx*) and (Mg,Fe)SiO$_3$ glass (*Pvt gl*) in a shock melt vein (*Vein*) in the Suizhou meteorite (Chen and Xie 2015)

Table 5.3 Chemical compositions of akimotoite and its host pyroxene in the Suizhou meteorite

Oxides	Akimotoite	Host pyroxene
FeO	13.71	13.76
MgO	27.79	27.86
MnO	0.45	0.47
CaO	0.89	0.86
K$_2$O	0.05	0.03
Na$_2$O	n.d.	n.d.
Al$_2$O$_3$	0.24	0.21
Cr$_2$O$_3$	0.16	0.18
TiO$_2$	0.16	0.19
SiO$_2$	56.55	56.44
Total	100	100

All data are measured by EDS in wt%
n.d. not detected

and interface-controlled growth similar to that of coarse-grained polycrystalline ringwoodite and majorite. Tomioka and Fujino (1997) interpreted the columnar texture as resulting from a martensitic-like mechanism. However, the TEM data presented by Tomioka and Fujino (1997) are also consistent with a coherent nucleation mechanism without martensitic-like shear (Sharp and DeCarli 2006).

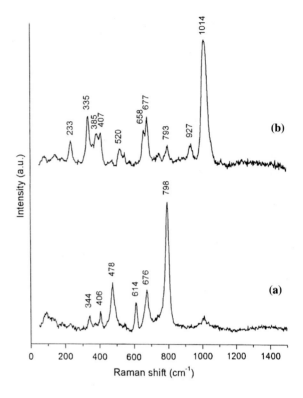

Fig. 5.11 Raman spectra of akimotoite (*a*) and host pyroxene (*b*) in the Suizhou meteorite (Chen and Xie 2015)

Miyagima et al. (2007) reported another mechanism of akimotoite formation in the shock melt veins of the Sixiangkou meteorite, namely the crystallization of akimotoite from shock-induced silicate melt. Akimotoite was identified in this meteorite by its characteristic plate-like morphology. The early crystallization of akimotoite from shock-induced silicate melt can be documented by its idiomorphic crystal morphology, absence of intergrowth textures with clinopyroxene, the coexistence with irregular crystals of majorite in the surrounding area, and higher contents of Al_2O_3 and Cr_2O_3 in akimotoite than clinopyroxene.

In the case of the Suizhou meteorite, akimotoite has two transformation mechanisms. Akimotoite that occurs along the fractures inside pyroxene fragments in the shock-produced melt veins and has the identical composition as low-Ca pyroxene forms from the solid-state transition of orthopyroxene. Akimotoite occurs in a granular texture where the crystal grains have a topotaxial relationship with the host pyroxene and forms a thin belt of randomly oriented akimotoite grains in between pyroxene and $(Mg,Fe)SiO_3$ glass belt which is in direct contact with a shock melt vein and is considered to be formed through homogeneous intracrystalline nucleation and interface-controlled growth similar to that of coarse-grained polycrystalline ringwoodite and majorite.

P-T Conditions

The occurrence of akimotoite in the Suizhou chondrite demonstrates that the aki-motoite forms at the region with high pressure and low temperature. It is well known from experiments that at high pressure and high temperature, pyroxene transforms to majorite. The transformation of enstatite to akimotoite requires temperatures in excess of 1550 °C at 22 GPa (Hogrefe et al. 1994). At much higher temperature, pyroxene becomes molten, and subsequently, majorite–pyrope garnet is crystallized from melt at high pressure and temperature. It is common to see the coexistence of coarse-grained polycrystalline ringwoodite aggregates and poly-crystalline majorite aggregates, and the high-pressure polymorphs of olivine and pyroxene formed from solid-state transitions, respectively, in the shock vein of meteorites including the Suizhou chondrite. However, akimotoite never occurs in contacting with ringwoodite. Obviously, the solid-state transitions both olivine to ringwoodite and pyroxene to majorite need both high pressure and high temperature.

5.2.4   Vitrified (Mg,Fe)SiO₃ Perovskite

Perovskite is a mineral with the composition $CaTiO_3$ and originally was thought to be cubic with Ca coordinated by 12 oxygens in a cubo-octahedral geometry and Ti in an octahedron. Further work showed that it is actually orthorhombic and that Ca is coordinated by eight oxygens. $(Mg,Fe)SiO_3$ perovskite, being one of the high-pressure polymorphs of pyroxene, is inferred to be the most abundant mineral phase in the Earth's lower mantle.

It is difficult to know who was the first scientist to realize the $MgSiO_3$ enstatite or Mg_2SiO_4 olivine which might transform to the perovskite structure at high pressure, but it was mentioned as a possibility by Ringwood (1962), and Ringwood and Major (1966) synthesized germanates with the orthorhombic perovskite structure, which was possible with an existing high-pressure apparatus. In the first successful experiment on a silicate, Liu (1974) obtained silicate perovskite by starting with pyrope $(Mg_3Al_2Si_3O_{12})$ and laser heating it in a diamond anvil cell at 27–32 GPa to produce $MgSiO_3$ perovskite plus corundum. Today, silicate perovskite is synthe-sized easily in diamond cells and in large-volume, multianvil apparatus at pressures of 22 GPa and above. Single crystal structure and elastic properties are determined by several investigators. There have, however, been arguments about the range of stability of perovskite at high temperatures and whether it might break down into MgO and SiO_2 under conditions existing in the lower mantle (Saxena et al. 1996). This was disputed by Mao et al. (1997) and Serghiou et al. (1998), and the exact conditions required for perovskite stability are still an open question (Prewitt and Downs 1998). Fei et al. (1996) presented a diagram showing that there is maximum amount of Fe^{2+} that $(Mg,Fe)SiO_3$ perovskite can accommodate at any given pressure and temperature. Mao et al. (1997) and McCammon (1998) found that $(Mg,Fe)SiO_3$ perovskite synthesized in both diamond cells and multianvil presses contain a significant amount of Fe^{3+} that can stabilize the structure. This also implies that Al^{3+} can have the same effect and the amount of Fe^{3+} and Al^{3+} in the lower mantle will have a strong influence on the range of stability of perovskite.

$(Mg,Fe)SiO_3$ perovskite has not been found in terrestrial rocks and thus has no mineral name. However, natural $(Mg,Fe)SiO_3$ perovskite was identified in shocked meteorites. Up to now, only two meteorites were reported to contain very small amount of perovskite. Tomioka and Fujino (1997) found that crystalline perovskite transformed from orthopyroxene in the Tenham meteorite. Sharp et al. (1997b) found vitrified perovskite in the shock melt veins of the Acfer 040 meteorite, which was interpreted as an amorphized perovskite during pressure release. The natural occurrence of perovskite is not only important for clarifying the phase transformations of pyroxene and the history of pressure and temperature in shocked meteorites, but also for understanding the phase transformation processes in the Earth's mantle.

In addition to high-pressure polymorphs, pyroxene may also be transformed to amorphous phase or glass at shock-produced high pressure and temperature. Pyroxene glass has been reported mainly on the TEM observations in some shocked meteorites, such as the retrograde glass inverted from majorite in the shock melt veins of Tenham meteorite (Price et al. 1979), the nanometer-sized pyroxene glass as a result of incipient shock melting in the ALH 84001 meteorite (Bell et al. 1999), and the Si–Al-rich glass and augite glass in shocked martian meteorites (Malavergne et al. 2001), and the $CaSiO_3$-rich glass is transformed from diopside in an H chondrite (Tomioka and Kimura 2003).

In comparison with previously found perovskite, vitrified perovskite, and pyroxene glass, in this section we describe a unique occurrence of fine-grained multi-granular $(Mg,Fe)SiO_3$ glass that intimately coexisted with majorite (Fig. 5.12) or akimotoite (Fig. 5.10) inside or contacting the Suizhou shock veins. We found large ovoid and zonal $(Mg,Fe)SiO_3$ glass grains that appear to represent the most extensive pyroxene–perovskite transformation observed in meteorites (Chen et al. 2004b).

Occurrence

Chen et al. (2004b) reported that within the shock melt veins in the Suizhou chondrite, some ovoid $(Mg,Fe)SiO_3$ glass grains with crust–core structure in which the grains are up to 120 μm in size. These $(Mg,Fe)SiO_3$ glass grains usually consist of a polycrystalline majorite rim surrounding an ovoid polygranular glassy silicate interior and are surrounded by the fine-grained matrix of garnet, ringwoodite, magnesiowüstite, FeNi metal, and troilite (Figs. 5.12 and 5.13). The majorite rim and the glassy interior have a very similar fine-grained structure. The glassy interior is made of subround or ovoid pockets ranging 60–90 μm in length and 30–70 μm in width. Abundant radiating fractures and cracks were observed in the majorite rim, whereby the cracks terminate at the surface of the glassy interior and the surrounding fine-grained matrix. High-magnification images revealed a granular texture of 1–3 μm in size in both the majorite rim and in the glassy phase. There is a

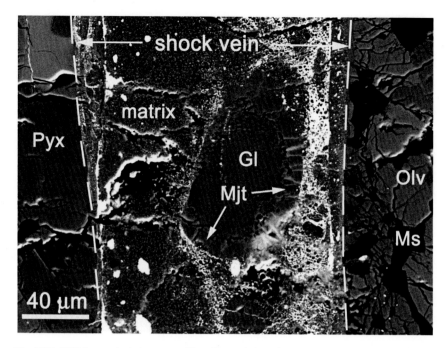

Fig. 5.12 BSE image depicting an ovoid grain consisting of a majorite rim (*Mjt*) and a (Mg,Fe) SiO_3 perovskite glassy interior (*Gl*) in the shock vein. The brighter material inside the shock vein is metal–troilite; Olv = olivine; Pyx = pyroxene; Ms = maskelynite (Chen et al. 2004b)

rough and uneven boundary between the majorite rim and the glassy interior. Small amounts of majorite inclusions of less than several micrometers in size occur in a narrow zone of 5–10 μm in width in the glassy interior adjacent to the inner wall of the majorite rim. Some majorite extends as branches from the inner wall of the majorite rim into the glassy interior (Figs. 5.12 and 5.13). It appears that there is a zone consisting of a mixture of two materials, i.e., majorite plus glass.

Our study strongly suggests that some precursor pyroxene grains inside the Suizhou shock melt veins were transformed to perovskite within the pyroxene grains due to a relatively low temperature, while at the rim region, pyroxene grains are transformed to majorite due to a higher temperature. After pressure release, perovskite is vitrified at post-shock temperature (Chen et al. 2004b).

Chemical Composition
Microprobe analyses show that the low-Ca pyroxene in the chondritic host, the perovskite glassy phase, and the majorite in the ovoid grains are identical in composition, especially the contents of Al_2O_3, CaO, Cr_2O_3, and Na_2O in these three phases are very low and very similar (Table 5.4). This means that the majorite rim in the ovoid grains and their glassy interior have the same composition as their precursor low-Ca pyroxene of the Suizhou meteorite, implying that there were no change in composition during the phase transition processes of pyroxene and no

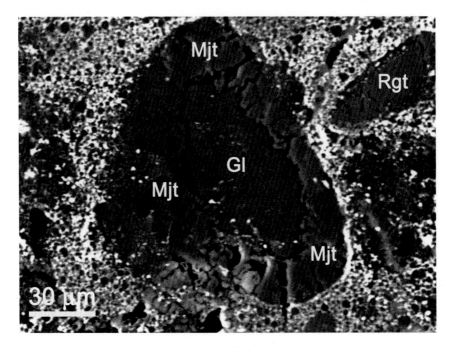

Fig. 5.13 BSE image showing another ovoid grain consisting of a majorite rim (*Mjt*) and a (Mg, Fe)SiO₃ perovskite glassy interior (*Gl*) in the Suizhou shock vein. The *brighter spots* in the glassy interior and the majorite rim are metal and troilite that could have been brought into pyroxene along fractures before the phase transition took place (Chen et al. 2004b)

Table 5.4 Electron microprobe analyses of (Mg,Fe)SiO₃ mineral phases (wt%)

Oxides	Low-Ca pyroxene (10)	Majorite (7)	Silicate glass (7)	Majorite–pyrope (8)
SiO₂	55.82	55.71	55.86	50.42
TiO₂	0.20	0.15	0.17	0.12
Al₂O₃	0.16	0.26	0.35	3.71
FeO	13.64	13.48	12.95	13.83
MnO	0.43	0.46	0.50	0.47
MgO	28.28	28.50	28.37	28.05
CaO	0.73	0.80	0.74	1.92
Na₂O	0.03	0.10	0.08	0.91
K₂O	0.01	0.01	0.03	0.05
Cr₂O₃	0.13	0.15	0.18	0.41
Total	99.43	99.62	99.23	99.79

The numbers in parenthesis are analysis number

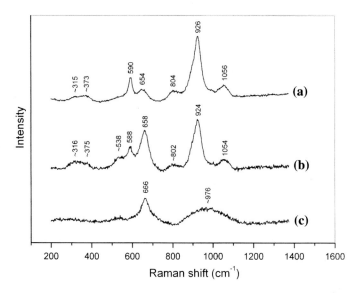

Fig. 5.14 Raman spectra of majorite that contains no glass core (*a*); majorite from the rim of an ovoid grain in Fig. 5.11 (*b*), and (Mg,Fe)SiO₃ perovskite glass from the interior of an ovoid grain in Fig. 5.11 (*c*) (Chen et al. 2004b)

exchanges in chemical elements with surrounding shock-induced silicate melt (Chen et al. 2004b).

Raman Spectroscopy

The Raman spectra of majorite and two (Mg,Fe)SiO₃ phases in an ovoid grain in Suizhou shock vein are shown in Fig. 5.14. The Raman spectrum of majorite that contains no glass core displays sharp bands at 926 and 690 cm^{-1} and weak bands at 315, 373, 654, 804, and 1056 cm^{-1} (Fig. 5.14a). The Raman spectrum of majorite in the rim of the ovoid grain displays sharp bands at 924 and 658 cm^{-1} and weak bands at 316, 375, 538, 588, 802, and 1054 cm^{-1} (Fig. 5.14b), whereas the Raman spectrum of the ovoid grain's glassy interior contains only two broad bands at 976 and 666 cm^{-1} which is typical for MgO–SiO₂ glasses (Fig. 5.14c), which could be attributed to Si–O stretching vibration of SiO₄ tetrahedra and to the inter-tetrahedra Si–O–Si vibration, respectively (McMillan 1984a, b).

Low-Ca Pyroxene–Perovskite Transformation Mechanism

So far, no crystalline (Mg,Fe)SiO₃ perovskite has been confirmed in natural assemblages. The best candidates for finding (Mg,Fe)SiO₃ perovskite are shock-metamorphosed meteorites. Sharp et al. (1997a, b) reported an amorphous (Mg,Fe)SiO₃ phase in association with akimotoite in the shock vein matrix of the Acfer 040 meteorite. These equant amorphous grains are rich in Al₂O₃, CaO, and Na₂O and are believed to have formed as crystalline phase originally from shock-induced dense melt. The formation of a amorphous (Mg,Fe)SiO₃ was interpreted to have amorphized from perovskite after pressure release. However,

Tomioka and Fujino (1997) reported the presence of $(Mg,Fe)SiO_3$ perovskite in the shock veins of the Tenham meteorite. The perovskite is found adjacent to some strongly deformed pyroxene fragments within the shock veins, showing that pyroxene was partially replaced by perovskite without melting. The TEM data presented by Tomioka and Fujino (1997) are consistent with a coherent nucleation mechanism without martensitic-like shear, and the pyroxene in Tenham meteorite is also partially transformed to a granular intergrowth of 200-nm $(Mg,Fe)SiO_3$ perovskite crystallites with the same chemical composition as the precursor pyroxene. This occurrence is consistent with homogeneous nucleation and interface growth.

The granular or equant texture of both the perovskite glass interior and the majorite rim of ovoid grains in Suizhou shock veins shows the material in the interior, and likewise, the majorite rim was a polycrystalline phase before the amorphization of the former. The similarity in the texture of the majorite rim and the amorphous interior strongly suggests a common origin during dynamic process. Furthermore, the $(Mg,Fe)SiO_3$ perovskite glass of the interior and the majorite in the rim have the same chemical composition as the precursor pyroxene. All these occurrences are indicative for a homogeneous nucleation and interface growth.

P-T Conditions

High-pressure experiments indicated that $MgSiO_3$ crystallizes in perovskite structures above 23 GPa and at ~ 2000 °C (Liu 1974, 1976; Gasparik 1992; Chen et al. 2004a). The finding of a large amount of $(Mg,Fe)SiO_3$ glassy phase, a vitrified perovskite, together with majorite has an important implication for phase transformation processes and the P-T history of shock veins.

5.2.5 *Lingunite, the Hollandite-Structured Plagioclase*

Hollandite is a structure type named after the mineral $BaMn_8O_{16}$. $NaAlSi_3O_8$ hollandite is the high-pressure polymorph of plagioclase which has the crystal structure of hollandite. The structure of hollandite is constructed of edge-sharing octahedra that form a four-sided eight-membered channel that is capable of containing a low-valence large-radius cation (Prewitt and Downs 1998). Together, the O atoms and the large cation form a hexagonal closely packed structure, a sort of analogue to the cubic closely packed perovskite. Ringwood et al. (1967) transformed sanidine into the hollandite structure with Al and Si randomly occupying the octahedral sites at 12 GPa and 900 °C. This was the first oxide structure identified with both Al and Si displaying sixfold coordination and only the second, after stishovite, with ^6Si. They postulated that feldspars are the most abundant minerals in the Earth' crust, and it is possible that the hollandite structure with ^6Al and ^6Si may be a common phase within the transition zone of the mantle. Liu (1978) reported the synthesis of $NaAlSi_3O_8$ hollandite at 21–24 GPa from jadeite plus stishovite. Above 24 GPa, the $NaAlSi_3O_8$ hollandite transforms to the calcium ferrite structure. Yamada et al. (1984) determined the crystal structure of the $KAlSi_3O_8$ phase by powder diffraction methods, and Zhang et al. (1993) reported the structure as a function of pressure.

First natural occurrence of $NaAlSi_3O_8$ hollandite was reported by Gillet et al. (2000). This high-pressure polymorph of plagioclase was found in the shock melt veins of a Chinese meteorite—Sixiangkou L6 chondrite—and occurs as intergrowths of $NaAlSi_3O_8$ hollandite + albitic glass. Most of the observed glass + hollandite intergrowths are surrounded by ringwoodite grains or the liquidus majorite–pyrope$_{SS}$ + magnesiowüstite, and in some rare cases, jadeitic pyroxene has also been found. The density of this new mineral is 3.80 gm/cm^{-1}, which is about 40 to 50 % higher than that of plagioclase (Gillet et al. 2000). The hollandite-structured plagioclase occurs in and adjacent to shock melt veins in meteorites. It consists of randomly oriented nanocrystals that range in size from 10 to about 100 nm. Optically, the hollandite-structured plagioclase is isotropic and looks like maskelynite (quenched dense plagioclase glass) or melted plagioclase (normal glass).

Another occurrence of $NaAlSi_3O_8$ hollandite in the shock veins of the Tenham chondrite was reported by Tomioka et al. (2000). It occurs as aggregates of extremely fine grains of several tens of nanometers. The hollandite phase seems to have formed from host feldspar in a solid-state reaction during the shock event.

During our recent micro-mineralogical studies on the shock melt veins of the Suizhou L6 chondrite, we also found quite a lot of $NaAlSi_3O_8$ hollandite grains in the shock melt veins of this meteorite (Xie et al. 2001b).

In 2004, the mineral hollandite-structured plagioclase was approved as a new mineral by the Committee of New Minerals, Nomenclatures and Classifications (CNMNC), and this new mineral was named as lingunite after the well-known experimental petrologist Liu, who first synthesized the $NaAlSi_3O_8$ hollandite in laboratory and conducted further characterization of this mineral (Liu 1978; Liu and El Goresy 2007).

The physical and chemical properties, as well as transformation mechanism of lingunite observed in shock melt veins of the Suizhou meteorite, are described in the following sections.

Occurrence

Lingunite, the $NaAlSi_3O_8$ hollandite, occurs as one of the major constituents of coarse-grained high-pressure mineral assemblage in the veins of Suizhou chondrite. The dark colored (under SEM) irregular or rounded grains of lingunite are smooth and unfractured with grain sizes ranging from 8 to 25 μm in diameter. No fractures among the other high-pressure polymorphs in the coarse-grained assemblage of the veins are ringwoodite, majorite, and tuite (Figs. 5.2 and 5.15). These high-pressure minerals are surrounded by the fine-grained majorite–pyrope$_{SS}$ + magnesiuwüstite and metal–troilite eutectic intergrowths.

Chemical Composition

The composition of hollandite-structured feldspars in shocked meteorites ranges from $KAlSi_3O_8$-rich (Langenhorst and Poirier 2000) to $NaAlSi_3O_8$-rich (Gillet et al.

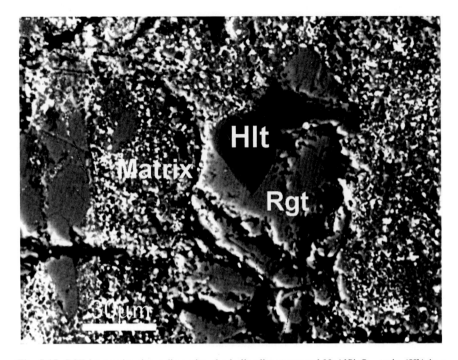

Fig. 5.15 BSE image showing a lingunite, the hollandite-structured $NaAlSi_3O_8$, grain (*Hlt*) in a Suizhou shock vein. Note the smooth and unfractured surface of the grain. Rgt = ringwoodite, Matrix = fine-grained vein matrix

2000) and intermediate plagioclase compositions (Langenhorst and Poirier 2000; Xie and Sharp 2004).

The chemical composition of lingunite in shock melt veins of Suizhou was determined by EPMA. Table 5.5 shows the chemical composition of lingunite and that of plagioclase outside the vein. From this table, we can see that lingunite in shock melt veins has almost the same chemical composition as that of plagioclase (oligoclase) outside the veins. This implies that the Suizhou lingunite was directly formed from plagioclase by phase transformation. The slightly higher FeO, MgO, and Na_2O contents and lower K_2O content in lingunite may be related to the influence of the shock-induced hot silicate melt in its surroundings.

Raman Spectroscopy

The Raman spectroscopic study of the lingunite in Suizhou melt veins and that of plagioclase outside the veins revealed that the Raman spectra of these two minerals are quite different (Fig. 5.16$_1$). The main peaks for plagioclase outside the veins are at 508 and 1110 cm^{-1} (Fig. 5.16$_3$). They correspond to Raman modes of Si–O–Si bending vibration and Si–O stretching of the SiO_4 tetrahedra, respectively, whereas the intense peak for lingunite is at 765 cm^{-1}, which is assigned to the Si–O stretching vibrations in the SiO_6 octahedra. The Raman spectrum of Suizhou

Table 5.5 Chemical composition of linguine in the Suizhou shock veins (wt%)

Oxides	SZ-Lglt-1	SZ-Lgt-2	SZ-Lgt-3	SZ-Lgt-4	Lingunite average	Plagioclase average
SiO_2	65.015	65.131	66.161	65.778	65.521	65.764
TiO_2	0.035	0.066	0.000	0.065	0.042	0.043
Al_2O_3	22.270	21.576	20.720	21.401	21.492	21.767
Cr_2O_3	0.000	0.011	0.000	0.022	0.008	0.027
MgO	0.145	0.886	0.506	0.965	0.625	0.004
CaO	2.139	2.454	2.211	2.059	2.216	2.209
MnO	0.000	0.047	0.047	0.000	0.025	0.017
FeO	1.091	1.176	0.925	1.127	1.080	0.414
BaO	0.000	0.000	0.000	0.000	0.000	0.018
Na_2O	9.300	8.401	9.302	9.533	9.134	8.867
K_2O	0.979	1.311	0.969	0.886	1.036	1.313
Total	100.974	101.059	100.841	101.836	101.179	100.443
An	83.59	81.82	83.34	84.71	83.37	80.94
Or	5.79	4.97	5.71	5.18	5.42	8.89
An	10.62	13.21	10.95	10.11	11.22	11.14
Cs	0.00	0.00	0.00	0.00	0.000	0.03

All values were determined by EPMA in wt%
Ab—Albite, Or—Orthoclase, An—Anorthite, Cs—Cesium

hollandite-structured $NaAlSi_3O_8$ can be compared with that of lingunite experimentally transformed from $KAlSi_3O_8$ feldspar (in the M'bale L6 meteorite) at 22 GPa and 1500 K (Fig. 5.16$_2$) and that of lingunite in Sixiangkou melt veins (Gillet et al. 2000). The lingunite in the Sixiangkou is interwoven with plagioclase glass. The lack of the broad band from 490 to 500 cm^{-1} and the less intense band near 1100 cm^{-1} in the grain of Suizhou lingunite indicates that no silicate glassy phase, such as albitic glass, occurs in the grain (Fig. 5.16$_1$). It is evident that this would be the first case of finding single-phase crystalline lingunite in natural materials.

Plagioclase–Lingunite Transformation Mechanism

In many heavily shocked chondrites, the transformation of plagioclase to the hollandite structure, like the olivine and pyroxene transformations mentioned in the last paragraphs, occurs in and adjacent to shock melt veins (Chen et al. 1996a, b; Gillet et al. 2000; Xie et al. 2001b). The origin of various hollandite-structured polymorphs of feldspars has been interpreted to be solid-state transformation (Tomioka et al. 2000; Langenhorst and Poirier 2000; Xie and Sharp 2004) and crystallization from melt (Gillet et al. 2000). The nanocrystalline granular texture is consistent with both origins if the melt was pure feldspar composition liquid (Sharp and DeCarli 2006). If formed by a solid-state mechanism, the microstructure suggests homogeneous nucleation and interface-controlled growth, as in the formation of ringwoodite. If the polycrystalline lingunites in melt veins are crystallized from feldspar composition liquids during shock compression, the feldspar composition

Fig. 5.16 Raman spectra of
lingunite (NaAlSi$_3$O$_8$
hollandite) in the Suizhou
melt veins (*1*), experimentally
produced KAlSi$_3$O$_8$-
hollandite (*2*) and plagioclase
outside the Suizhou veins (*3*)

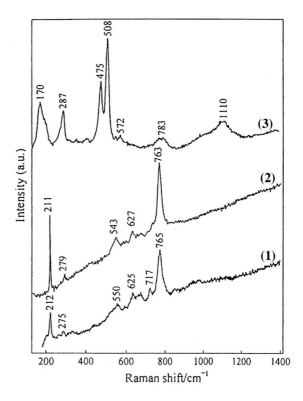

liquid did not mix with the surrounding chondritic liquid, as one would expect if the liquids are miscible. This suggests that most of the transformation is via a solid-state mechanism. However, veins of lingunite extending away from poly-crystalline aggregates in Sixiangkou L6 chondrite indicate that at least some of the plagioclase was molten during shock. The distinction between solid-state trans-formation and liquid crystallization is an important consideration when using the presence of lingunite to constrain shock pressures (Sharp and DeCarli 2006).

P-T Conditions

On the basis of high-pressure experiments, the presence of lingunite in Suizhou melt veins indicates that the pressure in the veins would not be higher than 23 GPa (Liu 1978). The occurrence of other coarse-grained and fine-grained high-pressure phases constraints the pressure in Suizhou melt veins to be in the range of 18–24 GPa and the temperatures between 1700 and 2300 °C (Xie et al. 2001a; Chen et al. 2003a). It is evident that the hollandite-structured plagioclase would be stable at high pressure up to 23–24 GPa and high temperature up to 1900–2000 °C. Such pressure and temperature regime corresponds to the pressure range from the lower part of mantle transition zone to the upper part of the lower mantle, e.g., 600–700 km deep in the Earth's mantle.

Geochemical Significance

Plagioclase is one of the most common rock-forming minerals in the Earth's crust. The discoveries of lingunite in natural materials, such as in Sixiangkou and Suizhou meteorites, are of important significance in understanding the Earth's mantle geochemistry, because this high-pressure phase is likely to form and survive in the transition zone and the upper part of the lower mantle (Liu 1978; Agee et al. 1995). In spite of its dense structure (40–50 % denser than plagioclase), the large four-sided tunnels along c-axis in the structure of this high-pressure phase might accommodate important large mono- and divalent cations, including Na, K, Rb, Sr, and Ba, and carry them from the Earth's surface down into the deep mantle during the subduction processes (Akaogi 2000). Some high-pressure partitioning experiments suggest that, when partial melting occurs in subducted crustal materials, hollandite-type plagioclase can preferentially incorporate several incompatible elements (K, Pb, Sr, light rare earth elements, and so forth) but is not likely to be a host for uranium and heavy rare earth elements (REE), relatively to the coexisting melt (Akaogi 2000). Therefore, the stability of lingunite will strongly influence trace element geochemistry of magma produced in the deep mantle as well as alkali transport processes in the transition zone and the lower mantle.

5.2.6 Tuite, the Dense Polymorph of Whitlockite

It is well known that whitlockite ($Ca_3(PO_4)_2$), together with apatite ($Ca_5(PO_4)_3(F, OH, Cl)$), is among the most important phosphate minerals found in lunar rocks, meteorites, and terrestrial rocks. The compound $Ca_3(PO_4)_2$ has four polymorphs including $\acute{\alpha}$-, α-, β-, and γ-phases, in which the β-phase (whitlockite) is stable at ambient conditions, the γ-phase is stable at high pressure, and the $\acute{\alpha}$- and α-phases are stable at high temperatures (Sugiyama and Tokonami 1987). The trigonal γ-$Ca_3(PO_4)_2$ is a dense phase isostructural with $Ba_3(PO_4)_2$ (Roux et al. 1978), which is 113 % denser than β-$Ca_3(PO_4)_2$. Thermodynamic computations indicate that whitlockite has a relatively low density making it unstable at higher pressures and that it transforms to γ-$Ca_3(PO_4)_2$ at pressures above 2.5 GPa at 1000 °C (Murayama et al. 1986). Multianvil high-pressure experiments using hydroxylapatite and fluorapatite as starting materials revealed that apatite decomposes to γ-$Ca_3(PO_4)_2$ at pressures above 12 GPa and temperatures from 1100 to 2300 °C (Murayama et al. 1986; Roux et al. 1978). Chen et al. (1995a, b) reported that chlorapatite in the shock veins of Sixiangkou meteorite was transformed to the phase A, an unknown high-pressure polymorph of phosphate mineral, at pressures about 23 GPa and temperatures about 2000 °C. However, no high-pressure polymorph of whitlockite was positively identified in any terrestrial and extraterrestrial rocks in the years of last century. Fortunately, the author of this book found the first natural occurrence of the high-pressure polymorph of whitlockite in a shock melt vein of the Suizhou meteorite. This new mineral and the mineral name "tuite" were approved by the Commission on New Minerals, Nomenclature and Classification (CNMNC) of the International Mineralogical Association (IMA) in 2002 (Xie et al. 2002, 2003). The occurrence, physical and chemical properties, and the formation mechanism of this new mineral are described in the following sections.

Occurrence

Tuite, the new high-pressure anhydrous phosphate phase, occurs in the Suizhou melt veins in the form of tabular polycrystalline grains with rounded outline in both tops. There are only four tuite grains that have been found in the Suizhou veins. The larger three grains are 20 μm × 10 μm, 18 μm × 15 μm, and 18 μm × 6 μm in sizes (Figs. 5.17, 5.18, and 5.19) and the smallest one is only 5 μm × 2.5 μm in size. The surface of tuite grains in the Suizhou vein is very smooth, and neither type showed any fractures, cleavages, or particular microstructure.

In Suizhou shock veins, tuite coexists with other high-pressure phases such as ringwoodite, majorite, and lingunite (Fig. 4.2). The grains of tuite are surrounded by fine-grained vein matrix consisted of majorite–pyrope garnet, FeNi metal, and troilite. Under reflected light and SEM in back-scattered electron mode, tuite grains show gray color that is close to that of ringwoodite and majorite, but lighter than lingunite. However, our synthetic tuite crystals are colorless and transparent with high refractive index and low birefringence. The luster of tuite is vitreous to resinous, and its steak is white.

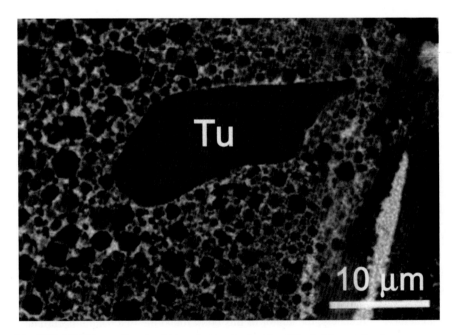

Fig. 5.17 BSE image of a tuite (*Tu*) grain in a shock vein of the Suizhou meteorite

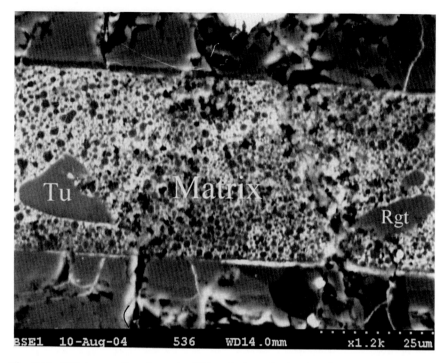

Fig. 5.18 BSE image of another tuite (*Tu*) grain in a shock vein of the Suizhou meteorite. Rgt = ringwoodite, Matrix = fine-grained vein matrix

Optical Properties

Tuite cannot be identified through an optical microscope in transmitted light due to its very small size and thickness and very low abundance. Furthermore, the presence of matrix minerals (FeNi metal, Fe-sulfide, and garnet) underneath the tuite grains in a 30-μm-thick polished thin section makes these tuite grains (about 5 μm in thickness) almost "opaque." Therefore, the optical properties of tuite were determined on the colorless and transparent synthetic tuite crystals of 60–130 μm in size.

The results of our measurements for synthetic tuite crystals are as follows: uniaxial (+), $\varepsilon = 1.706$ (3) and $\omega = 1.701$ (4). Our results show that the optical properties of tuite are in good consistence with the description of Murayama et al. (1986) for their synthetic high-pressure $Ca_3(PO_4)_2$ phase that shows high refractive indices and low birefringence and are markedly different from those of natural whitlockite from Palermo, New Hampshire, USA, which has chemical composition (CaO-46.90, MgO-2.53, P_2O_5-45.68 wt%) similar to that of our tuite in Suizhou but shows uniaxial (−) optical sign with refractive indices of $\omega = 1.629$ and $\varepsilon = 1.626$ (Frondel 1941).

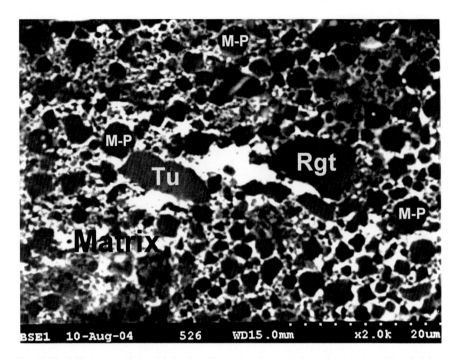

Fig. 5.19 BSE image of the third tuite (*Tu*) grain observed in a shock vein of the Suizhou meteorite. Rgt = ringwoodite, M-P = majorite–pyrope garnet in solid solution, Matrix = fine-grained vein matrix

Chemical Composition

Chemical composition of tuite in the Suizhou meteorite was obtained with a Cameca SX-51 electron probe micro-analyzer. The tabular tuite grain of 10×20 μm in size on a polished thin section was probed in three areas (Table 5.6). Results show that this mineral appears homogeneous with empirical formula $Ca_{17.55}(Mg_{1.89}, Fe_{0.11})_{2.00}(Na_{1.93}, K_{0.02})_{1.96}(P_{1.01}O_4)_{14}$, or $(Ca_{2.52}Mg_{0.26})_{2.78}Na_{0.25}(P_{1.02}O_4)_2$ which is very similar to that of whitlockite outside the vein, and there are no additional elements incorporated with this high-pressure phosphate phase from the surrounding matrix melt, and no chlorine was detected. Hence, tuite is thought to have been transformed directly from original whitlockite during a shock event and not produced from decomposition of chlorapatite. The simplified formula for tuite may be given as $(Ca,Mg,Na)_3(PO_4)_2$ or $Ca_3(PO_4)_2$.

Raman Spectroscopy

The Raman spectra for the 10 μm × 20 μm grain of tuite in a Suizhou shock vein are shown in Fig. 5.20. They were conducted on the same 3 points of this tuite grain where EPMA data were measured. Interestingly, all three obtained spectra are almost identical and show only one intense peak at 974–975 cm^{-1} with a shoulder at 997 cm^{-1}, two less intense peaks at 411 and 577–578 cm^{-1}, and three weak peaks at 191–192 cm^{-1}, 640–641 cm^{-1}, and 1095 cm^{-1}. The intense Raman peak at

Table 5.6 Electron microprobe analyses of tuite and whitlockite (wt%)

	Tuite				Whitlockite
	1	2	3	Average	
TiO_2	0.06	0.03	0.04	0.04	0.06
FeO	0.33	0.44	0.37	0.38	0.28
MgO	3.60	3.55	3.59	3.58	3.27
CaO	45.91	46.41	46.10	46.14	46.62
NiO	0.03	0.09	0.03	0.05	0.08
Na_2O	2.78	2.82	2.81	2.80	2.57
K_2O	0.05	0.09	0.07	0.07	0.03
Cr_2O_3	0.00	0.01	0.00	0.00	0.03
P_2O_5	47.14	47.19	47.15	47.16	47.67
Total	99.90	100.63	100.16	100.22	100.61

Formula of tuite is given as follows: $Ca_{17.55}(Mg_{1.89},Fe_{0.11})_{2.00}(Na_{1.93},K_{0.03})_{1.96}(P_{1.01}O_4)_{14}$, or $(Ca_{2.51},Mg_{0.29})_{2.80}Na_{0.28}(P_{1.01}O_4)_2$

Formula of whitlockite is given as follows: $Ca_{17.63}(Mg_{1.72},Fe_{0.08})_{1.80}(Na_{1.76},K_{0.02})_{1.78}(P_{1.02}O_4)_{14}$, or $(Ca_{2.52}Mg_{0.26})_{2.78}Na_{0.25}(P_{1.02}O_4)_2$

The calculation bases for both formulae are $O = 56$ and $O = 8$, respectively

Fig. 5.20 Raman spectra of the Suizhou tuite measured in three spots (SZ-V-2-2 to SZ-V-2-4) (Xie et al. 2003)

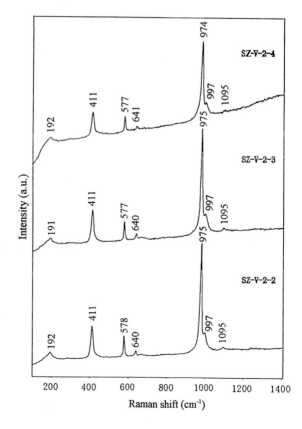

Fig. 5.21 Comparison of
Raman spectra of Suizhou
whitlockite (merrillite) (*a*),
γ-Ca$_3$(PO$_4$)$_3$ (tuite) in Suizhou
veins (*b*), and synthesized
γ-Ca$_3$(PO$_4$)$_3$ (*c*) (Xie et al.
2002)

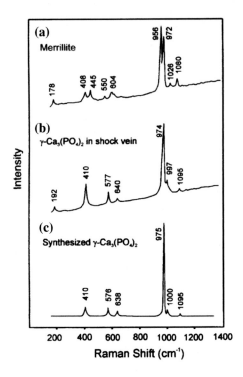

974–975 cm^{-1} can be assigned to the $v1$ symmetric stretching vibration of PO$_4$
group, and the peak at 1095 cm^{-1} corresponds to $v3$ asymmetric stretching vibration of PO$_4$, whereas those at 577–578 cm^{-1} and 640–641 cm^{-1} are associated with
$v4$ bending mode and those less than 411 cm^{-1} with the lattice modes.

For comparison, the Raman spectra of Suizhou whitlockite, natural γ-Ca$_3$(PO$_4$)$_3$
(tuite) in Suizhou veins, and synthesized γ-Ca$_3$(PO$_4$)$_3$ are shown in Fig. 5.21. Here,
we can see that the Raman spectra of natural and synthetic γ-Ca$_3$(PO$_4$)$_3$ are identical, different from that of whitlockite.

It should be pointed out that the Raman spectra of tuite display only one strong
peak at 974–975 cm^{-1} that is quite different from the strong doublet of REE-poor
whitlockite from 950 to 976 cm^{-1} (Chen et al. 1995a; Jolliff et al. 1996). The
well-resolved strong peak of tuite at 974 to −975 cm^{-1} is also distinct from the
asymmetric single peak or the very poorly resolved doublet from REE-rich whitlockite (Jolliff et al. 1996). Since both tuite in Suizhou veins and whitlockite outside
the Suizhou veins have the same REE-poor composition, and the shock vein is full
of high-pressure minerals, the compositional and structural data suggest that a phase
transformation from whitlockite to its high-pressure polymorph did take place in the
Suizhou shock veins.

X-ray Diffraction Data of Synthetic Tuite
Natural tuite grains we found in the Suizhou shock veins are extremely small. The
larger one is 20 μm × 10 μm, and the smaller one is only 5 μm × 2.5 μm in sizes.

These grains are surrounded all round and underneath by the vein matrix minerals, namely majorite garnet, FeNi metal, and troilite (Fig. 5.17). It is impossible to obtain X-ray diffraction pattern from such complicated mineral samples. So, it is important to find X-ray diffraction data of synthetic tuite first. However, Murayama et al. (1986) stated that they examined the synthetic γ-$Ca_3(PO_4)_2$ phase by X-ray powder diffraction analysis, but no diffraction data of this phase have been published in the literature. Therefore, we had to synthesize the γ-$Ca_3(PO_4)_2$ phase in the laboratory at 14 GPa and 1400 °C using $Ca_3(PO_4)_2$ powder as starting material and a multianvil apparatus at the Geophysical Laboratory of the Carnegie Institution of Washington. Then, we collected powder X-ray diffraction data for our synthetic tuite with a Rigaku PSPC-MDG2000 X-ray micro-diffractometer equipped with an imaging plate detector system. All collected diffraction peaks of synthetic tuite are sharp.

The recorded powder X-ray diffraction lines are shown in Fig. 5.22, and the data are listed in Table 5.7. The X-ray diffraction data reveal that the synthetic tuite has a trigonal unit cell with cell parameters of $a = 5.2576$ and $c = 18.7049$, and density (calc.) = 3.452 (g/cm^3) (Table 5.8). Our synthetic tuite phase is identical to the high-pressure γ-$Ca_3(PO_4)_2$ phase found by Maruyama et al. (1986) as the decomposition product of apatite at high pressures and temperatures. The structure refinements are also based on crystal structural model for the γ-$Ca_3(PO_4)_2$ phase proposed by Sugiyama and Tokonami (1987). The calculated d-values agree very well with the observed X-ray diffraction data (Xie et al. 2002).

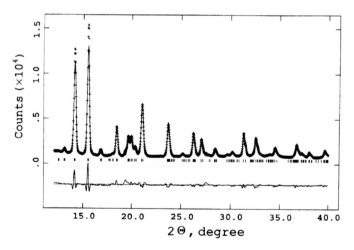

Fig. 5.22 Observed (*crosses*) and calculated (*solid line*) X-ray diffraction pattern for the synthesized γ-$Ca_3(PO_4)_3$ high-pressure phase. The observed data were collected with a wavelength of 0.70927 Å (Mo). *Tick marks* for peak positions of the high-pressure phase are shown below the pattern. The *difference curve* is shown at the *bottom*. The refinement is based on the space group R-3 m with trigonal cell parameters, $a = 5.2576$ (2) Å and $c = 18.7049$ (13) Å (Xie et al. 2002)

Table 5.7 X-ray powder diffraction data of synthetic γ-Ca₃(PO₄)₂ and tuite

h k l	Synthetic γ-Ca$_3$(PO$_4$)$_2$		Tuite	
	$d_{meas.}$	I/I_0	$d_{meas.}$	I/I_0
1 0 1	4.4240	1.6	4.428	6.5
0 0 6	3.1175	4.0	3.118	5.7
1 0 5	2.8905	83	2.891	80.3
1 1 0	2.6288	100	2.628	100
1 1 3	2.4223	4.7	2.423	4.6
2 0 2	2.2120	24.3	2.214	20.3
1 0 8	2.0799	6.2	2.080	5.0
0 0 9	2.0783	10.5	2.078	12.2
2 0 4	2.0469	16.1	2.047	15.7
1 1 6	2.0097	8.2	2.009	8.0
2 0 5	1.9448	50.0	1.945	47.3
2 0 7	1.7324	3.0	1.734	4.3
0 1 10	1.7302	30.8	1.730	24.5
1 1 9	1.6303	3.1	1.627	5.9
1 2 5	1.5634	23.2	1.567	22.2
3 0 0	1.5177	18.2	1.518	18.6
2 0 10	1.4452	8.8	1.445	10.7
1 2 8	1.3860	1.7		
3 0 6	1.3646	1.6		
0 2 11	1.3624	4.4		
2 2 0	1.3144	27.0		
1 2 10	1.2665	23.9		
0 0 15	1.2470	1.2		
1 3 5	1.1965	11.9		
4 0 2	1.1300	1.4		
1 1 15	1.1267	17		
2 2 9	1.1109	4.9		
4 0 4	1.1060	1.2		
4 0 5	1.0890	6.7		
1 3 10	1.0466	10.6		
2 3 5	1.0061	7.4		
1 4 0	0.9936	8.5		
0 4 10	0.9724	2.4		
0 3 15	0.9635	9.5		
1 0 19	0.9622	1.2		
1 0 20	0.9161	2.0		
2 3 10	0.9120	5.7		
2 2 15	0.9046	2.6		
5 0 5	0.8848	1.5		

Table 5.8 Cell parameters of synthesized and natural polymorphs of $Ca_3(PO_4)_2$

Parameters	Synthesized γ-$Ca_3(PO_4)_2$	Synthesized γ-$Ca_3(PO_4)_2$	Natural γ-$Ca_3(PO_4)_2$	Natural β-$Ca_3(PO_4)_2$
a (Å)	5.2487 (6)	5.2576 (2)	5.258 (1)	10.37
c (Å)	18.6735 (36)	18.7049 (13)	18.727 (3)	37.19
Volume (Å^3)	445.5 (1)	447.7 (6)	448.3 (6)	
Composition	$Ca_3(PO_4)_2$	$Ca_3(PO_4)_2$	$Ca_3(PO_4)_2$	$Ca_3(PO_4)_2$
Z	3	3	3	21
Density (calc.) (g/cm^3)	3.469	3.452	3.447	3.12
Density (meas.) (g/cm^3)	3.462	–	–	3.12
Source	Sugiyama and Tokonami (1987)	Xie et al. (2002)	Xie et al. (2002)	Gopal (1974)

Synchrotron Radiation X-ray Diffraction Data of Natural Tuite

The EPMA and Raman spectroscopic investigations revealed that natural tuite has identical chemical composition of whitlockite, but its crystal structure is different with that of whitlockite. Therefore, it is important to obtain its X-ray diffraction data for comparing with those obtained from the synthetic tuite and for characterization of this mineral. For this reason, the larger tuite grain in a thin section was chosen for synchrotron radiation X-ray diffraction in situ analysis.

The X-ray beam of 15 μm × 15 μm in size was focused on the tuite grain of 10 μm × 20 μm in size (Fig. 5.17). The X-ray patterns in fixed orientation of sample indicate polycrystalline nature of the grain, and the collected diffraction lines came not only from the tuite grain but also from all three fine-grained matrix minerals surrounding and underneath the tuite grain. Since the X-ray diffraction patterns of these three matrix minerals are well known, we were able to find a set of lines that do not belong to these minerals but to tuite. Finally, 17 diffraction lines were collected from this natural tuite grain (Xie et al. 2003).

The synchrotron radiation X-ray diffraction lines of natural tuite were compared with the observed X-ray diffraction lines of synthetic tuite. It was found that all reflections ($h\,k\,l$ = 105, 110, 113, 113, 202, 204, 116, 205, 10 10, 125, 300, etc.) and some of their overtones (101, 220, 20 10, etc.) on the diagram of synthetic tuite can be compared with the diffraction lines we collected from natural tuite (Fig. 5.23). Therefore, the powder diffraction pattern of 17 lines with d-values, intensities (I), relative intensities (I/I_o), and Miller indices ($h\,k\,l$) for natural tuite was successfully obtained (Table 5.7).

The trigonal unit cell parameters for natural tuite in the Suizhou meteorite were then calculated as follows: a = 5.258 (1) Å and c = 18.727 (3) Å; space group R-3 m; and density (calc.) = 3.447 (g/cm^3), where the numbers in parentheses are standard deviations in the last significant digits. The structure refinements are also based on crystal structural model for the γ-$Ca_3(PO_4)_2$ phase proposed by Sugiyama and Tokonami (1987).

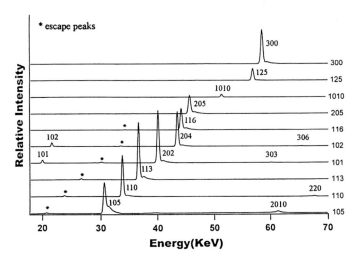

Fig. 5.23 SRXRD patterns of synthetic tuite with randomly picked 10 reflections (Xie et al. 2002)

Whitlockite–Tuite Transformation Mechanism

It was revealed that the tuite grains in the shock veins of the Suizhou chondrite occur as polycrystalline aggregates that have the same chemical composition as the whitlockite in the unmelted chondritic rock of the same chondrite (Xie et al. 2002, 2003). This implies that tuite formed not through decomposition of apatite as it was revealed by experiments of Murayama et al. (1986), but from its host mineral whitlockite via an isochemical solid-state transformation mechanism during shock compression (Xie et al. 2003, 2005a, b).

The polycrystalline nature of the Suizhou tuite grains and the constant composition for three different tuite grains indicate the homogeneous features in chemical composition of tuite crystallites within grains. Therefore, it is reasonable to assume that whitlockite–tuite transformation mechanism in the Suizhou meteorite can be interpreted to be the homogeneous intracrystalline nucleation of tuite crystallites throughout the host whitlockite grain and followed by the interface-controlled growth of crystallites in the grain.

P-T Conditions

It is revealed that shock-produced *P-T* conditions in Suizhou veins were adequate to result in the formation of a series of high-pressure phases. Phase transition from plagioclase to lingunite constrains the pressure and temperature in the shock vein to about 23 GPa and 2000 °C (Liu 1978; Yagi et al. 1994; Gillett et al. 2000; Xie et al. 2001b). Such *P-T* conditions developed in the veins are also available for the solid-state phase transformations from olivine to ringwoodite and from pyroxene to majorite (Chen et al. 1996a; Xie et al. 2001b). Even though the phase diagrams in Murayama et al. (1986) indicate that the γ-$Ca_3(PO_4)_2$ might form at as low as 12 GPa, it would still be present at 20 GPa or higher. The P-T condition suggested by tuite is consistent with the P-T conditions (up to 20–23 GPa and 1900–2000 °C) indicated by the other high-pressure phases in Suizhou veins.

Like the Suizhou meteorite, the shock veins of Sixiangkou meteorite contain abundant high-pressure minerals including ringwoodite, majorite, lingunite, fine-grained majorite–pyrope garnet, and magnesiowüstite, which set an upper bound for P-T regime of shock vein to be 20–24 GPa and $\sim$2000 °C (Chen et al. 1996a; Gillet et al. 2000). One may expect that the whitlockite in the shock veins of Sixiangkou should have been also transformed into its high-pressure polymorphs. However, no phase transitions from whitlockite to γ-$Ca_3(PO_4)_2$ phase are observed from its shock veins (Chen et al. 1995a, b). Experiments indicate that the transformation from whitlockite to γ-$Ca_3(PO_4)_2$ takes place at a pressure lower than 20 GPa (Murayama et al. 1986; Sugiyama and Tokonami 1987). It, therefore, shows that the shock-induced pressures in the shock veins of both Suizhou and Sixiangkou are sufficient for the phase transition from whitlockite to γ-$Ca_3(PO_4)_2$. Considering that the Sixiangkou was much more severely shock-metamorphosed than the Suizhou and that there is a great difference in the abundance and thickness of shock veins between the Suizhou and the Sixiangkou, we explain that their cooling history could play a key role in the quenching of this high-pressure polymorph of whitlockite. Evidently, the extremely thin shock veins in the Suizhou meteorite reserve nearly all shock-induced high-pressure phases, whereas an back transformation from the γ-$Ca_3(PO_4)_2$ to whitlockite should have taken place in the shock veins of the Sixiangkou meteorite. In fact, Chen et al. (1995a, b) have observed a difference in Raman spectra between the intact whitlockite in the chondritic region and the whitlockite in the shock veins of Sixiangkou, which could be a hint for the back transformation from the γ-$Ca_3(PO_4)_2$ phase to whitlockite.

Geochemical Significance
Tuite, as the dense phase of whitlockite, has a larger density in comparison with the common minerals in the Earth's mantle, such as majorite garnet and the spinel phase of ferromagnesian orthosilicates, and may stably exist at the Earth's deep horizon. The γ-$Ca_3(PO_4)_2$-structured phases, including tuite, have 12-coordinated Ca(1) site and 10-coordinated Ca(2) site with mean M-O distances of 2.74 and 2.59 Å, respectively (Sugiyama and Tokonami 1987). These bonds are much longer than those of 8-coordinated Ca site in garnet and diopside (M-O $\approx$ 2.40 Å). Consequently, the γ-$Ca_3(PO_4)_2$-structured phases have the potential to accommodate very large lithophile elements (VLLE), such as Sr and Ba, and REE, more easily than common mantle minerals and persist as stable crystalline phases in the whole upper mantle (Beswick and Carmichael 1978; Murayama et al. 1986). Therefore, we have identified at least one important natural crystalline phase that is expected to function as a reservoir of VLLE and REE in the deep upper mantle.

5.2.7 High-Pressure Phase of Chlorapatite Decomposition

Apatite [$Ca_5(PO_4)_3(OH,F,Cl)$] and whitlockite [$Ca_3(PO_4)_2$] are among the most important phosphate minerals found in terrestrial rocks, lunar samples, and meteorites (Griffin et al. 1972; Prewitt 1975; Dowty 1977; Buchwald 1984; Nash 1984; Rubin 1997; Engvik et al. 2009). It is known that apatite occurring in meteorites is mainly chlorapatite [$Ca_5(PO_4)_3Cl$]. Chen et al. (1995a, b) reported that chlorapatite in shock veins of the Sixiangkou meteorite was transformed to an unknown high-pressure polymorph called phase A. Subsequently, Xie et al. (2002, 2003) found the high-pressure polymorph of whitlockite with the structure of γ-$Ca_3(PO_4)_2$ in shock veins of the Suizhou meteorite and named it tuite. This section reports our discovery of tuite from a shock vein of the same Suizhou meteorite, which is believed to have formed through the decomposition of chlorapatite, rather than phase transformation from whitlockite (Xie et al. 2013).

Such a decomposition process has also been verified by our high-pressure experiment using synthetic chlorapatite as starting material at 15 GPa and 1573 K (Xie et al. 2013). By comparing the EPMA data and Raman spectra from our natural and synthetic tuites with those for phase A found in shock veins of the Sixiangkou meteorite by Chen et al. (1995a, b), we conclude that phase A is actually a product of incomplete decomposition of chlorapatite. It is revealed that phase A is different in some extend from our tuite phase formed from chlorapatite decomposition both in chemical composition and in crystal structure. Hence, we assume that these two tuite grains in the Suizhou vein can be considered as the high-pressure phase of chlorapatite decomposition.

Occurrence

In a shock vein of the Suizhou meteorite, we detected three Ca-phosphate grains from the EDS measurements (Fig. 5.24). The first smaller one (labeled T1 in Fig. 5.25) is $\sim$10 μm $\times$ 20 μm, the second larger one (labeled T2 in Fig. 5.25) $\sim$20 μm $\times$ 35 μm, and the third larger one (labeled T3 in Fig. 5.26) $\sim$12 μm $\times$ 20 μm. These three grains exhibit no cracks or fractures and are gray in color under reflected light, looking like other high-pressure silicate minerals, such as ringwoodite and majorite in the vein. However, our EDS study revealed that they are not silicate but phosphate of calcium, and their Raman spectra (Fig. 5.27a, b) are almost identical to that of tuite (γ-$Ca_3(PO_4)_2$), the high-pressure polymorph of whitlockite (Fig. 5.20).

Chemical Composition

The EPMA results reveal that, in addition to the major components CaO and P_2O_5, the T1 and T2 grains contain minor amounts of Na_2O and MgO (Table 5.9). Most remarkably, these two grains also contain small amounts of Cl (0.15 wt% in T1 and 1.65 wt% in T2). The marked difference in the Cl content between T1 and T2 leads us to conclude that grain T1 represents a complete transformation product of chlorapatite to tuite, whereas grain T2, though mostly tuite, is likely to contain some relics of the precursor chlorapatite.

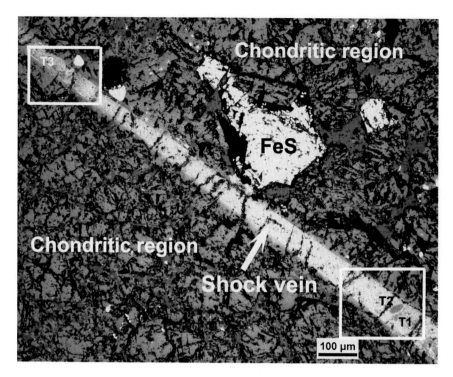

Fig. 5.24 Reflected light microphotograph showing a Suizhou shock vein containing two tuite grains *T1* and *T2* inside the *lower right rectangle* and another tuite grain *T3* inside the *upper left rectangle area* (Xie et al. 2013)

It is interesting to mention that the observed third phosphate grain (T3) occurs in the same shock vein where tuite grains T1 and T2 occurred (Fig. 5.24). On the basis of electron microprobe analysis (Table 5.9, last column) and Raman spectroscopy, the T3 grain is identified as tuite as well, but its precursor mineral is whitlockite, rather than chlorapatite, because it contains 2.88 wt% of Na_2O and 3.43 wt% of MgO, and no detectable Cl.

Raman Spectroscopy

The Raman spectra of the T1 and T2 grains (Fig. 5.27a, b) are directly comparable to the spectrum of tuite transformed from whitlockite at high pressure (Fig. 5.20). Specifically, both Raman spectra of the T1 and T2 grains display an intense peak at 977 cm^{-1}; three less intense peaks at 1096, 412, and 577 cm^{-1}; and two weak peaks at 1001–1002 and 640 cm^{-1}, which are characteristic for Raman spectrum of tuite.

On the basis of Raman spectroscopy and EPMA data, we conclude that the grain T1 has completely transformed to tuite, whereas grain T2, though mostly tuite, is likely to contain some relics of the precursor chlorapatite. In fact, the very weak peak at 957 cm^{-1} on the Raman spectrum for the T2 grain (Fig. 5.27b) may lend further support to our conclusion, because this peak does not belong to tuite, but is characteristic of chlorapatite.

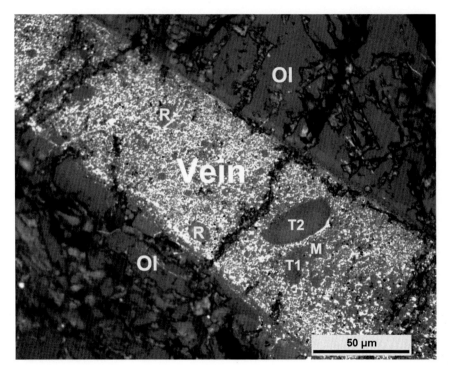

Fig. 5.25 Enlarged reflected light microphotograph of the *lower right rectangle area* in Fig. 5.4, showing the two tuite grains *T1* and *T2*. R = ringwoodite, M = majorite, Ol = olivine (Xie et al. 2013)

Synthesized High-Pressure Phase Tuite from Chlorapatite

The phase transition of chlorapatite to tuite was conducted using synthetic chlorapatite as starting material at high-pressure and high-temperature condition. First, reagent-grade $CaCO_3$, NH_4Cl, and $NH_4H_2PO_4$ powders were mixed in a proportion corresponding to the $Ca_5(PO_4)_3Cl$ stoichiometry. The mixture was ground for 2 h in an agate mortar and pressed into pellets with a diameter of 5 mm under uniaxial pressure of 30 MPa. Second, the pellets were sintered in a conventional muffle furnace at 1273 K for 36 h to form a single phase of $Ca_5(PO_4)_3Cl$, which was confirmed by powder X-ray diffractometer. Third, the $Ca_5(PO_4)_3Cl$ was used as the starting material and put into a Pt capsule to synthesis tuite at 15 GPa and 1573 K for 24 h using a multianvil apparatus. The high-pressure and high-temperature experimental details were described in Xue et al. (2009). Our high-pressure synthesized semi-euhedral tuite crystals up to 50 μm in size are shown in Fig. 5.28.

The electron microprobe analyses on ten synthesized tuite crystals produced an average composition (wt%): CaO—54.24, P_2O_5—46.23, total—100.48, yielding a chemical formula of $Ca_3(PO_4)_2$ when normalized on the basis of 8 O atoms. The Raman spectrum of synthetic tuite is given in Fig. 5.27c, which consists of one intense peak at 977 cm^{-1}; two less intense peaks at 412 and 1095 cm^{-1}; and three

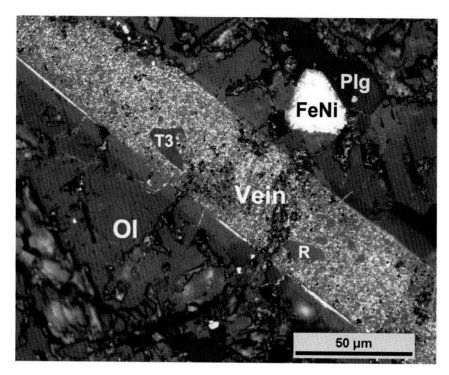

Fig. 5.26 Enlarged reflected light microphotograph of the *upper left rectangle area* in Fig. 5.4, showing the tuite grain *T*. R = ringwoodite, Ol = olivine, Plg = plagioclase, FeNi = metal (Xie et al. 2013)

weak peaks at 578, 640, and 1001 cm^{-1}. This spectrum is identical to the Raman spectra of Suizhou tuite, and the γ-Ca$_3$(PO$_4$)$_3$ phase synthesized directly from Ca$_3$(PO$_4$)$_2$ at 14 GPa and 1400 °C for 24 h (Fig. 5.21c). Hence, we conclude that the high-pressure phase of chlorapatite decomposition is tuite.

X-ray Diffraction Data of Synthetic Tuite from Chlorapatite Decomposition
We also measured the X-ray powder diffraction patterns for both synthetic tuite and its starting material chlorapatite (Fig. 5.29). The calculated cell parameters of the synthetic tuite are as follows: a = 5.2507 (13) Å, c = 18.6746 (60) Å, V = 445.89 (21) Å^3, and ρ = 3.465 (2) g/cm^3. From the above experimental data, it is evident that we have successfully synthesized tuite from chlorapatite through a decomposition reaction.

It should be pointed out that we did not detect any other phase except tuite in the synthesized sample by means of both Raman spectroscopy and X-ray diffraction. The microprobe element maps for the synthesized tuite sample show only Ca and P, and no measurable Cl. Nonetheless, very low level of Cl was detected in some interstices between tuite crystals (Fig. 5.30). The EDS analyses of elements Ca, P, O, and Cl also show that the contents of Cl for six synthesized tuite crystals with different sizes are extremely low (0.01–0.08 wt%) (Fig. 5.31 and Table 5.10).

Fig. 5.27 Raman spectra of natural and synthetic tuites transformed from chlorapatite decomposition. *a*—Suizhou tuite grain T1, *b*—Suizhou tuite grain T2, *c*—tuite synthesized from chlorapatite (Xie et al. 2013)

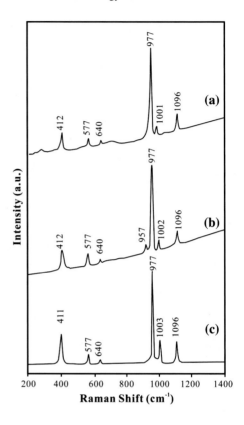

Table 5.9 Microprobe analyses of three tuite grains in a Suizhou shock vein (wt%)

	Grain	T1	Grain	T2	Grain	T3
	(3)	s.d.	(3)	s.d.	(2)	s.d.
Na_2O	0.15	0.01	0.10	0.03	2.88	0.12
K_2O	0.01	0.01	–	–	–	–
MgO	0.85	0.12	0.24	0.07	3.43	0.25
CaO	53.01	0.08	54.35	0.41	46.71	0.58
MnO	–	–	0.01	0.01	–	–
Al_2O_3	–	–	–	–	–	
P_2O_5	45.24	0.07	44.65	0.54	47.90	0.43
Cl	0.15	0.02	1.65	0.14	–	–
	99.40		101.01			
–O = Cl	0.03		0.44			
Total	99.37		100.57		100.92	

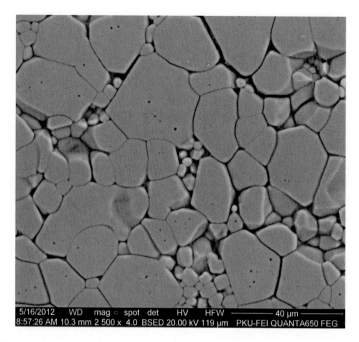

Fig. 5.28 BSE image showing aggregate of semi-euhedral tuite crystals synthesized from chlorapatite at 15 GPa and 1573 K for 24 h using a multianvil apparatus. Note the different sizes of tuite crystals (Xie et al. 2013)

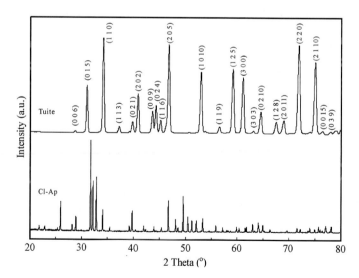

Fig. 5.29 X-ray powder diffraction patterns of synthesized tuite (*Tuite*) and starting material chlorapatite (*Cl-Ap*). Note the big difference between these two patterns (Xie et al. 2013)

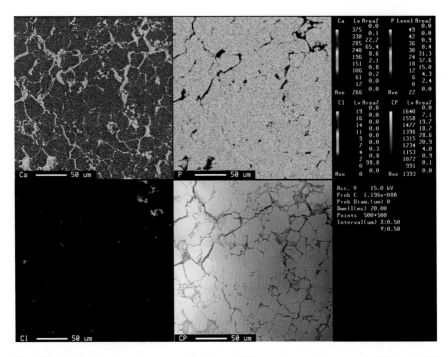

Fig. 5.30 Microprobe element maps of Ca, P, and Cl for synthesized tuite crystals. CP is the BSE image of synthesized tuite sample. Note that the chlorine is absent in tuite crystals (Xie et al. 2013)

Chlorapatite–Tuite Transformation Mechanism

Tuite was first identified in very thin shock veins of the Suizhou L6 chondrite as the high-pressure polymorph of whitlockite (Xie et al. 2002, 2003). Our recent study further demonstrates that tuite can also be formed as the high-pressure product of chlorapatite decomposition.

It is easy to understand that whitlockite changes to tuite just through an iso-chemical solid-state phase transition under high pressures and high temperatures, but the change of chlorapatite to tuite requires a decomposition reaction: $2 \text{ Ca}_5(\text{PO}_4)_3\text{Cl} \rightarrow 3 \text{ } \gamma\text{-Ca}_3(\text{PO}_4)_2 + \text{CaCl}_2$. Hence, the search for the CaCl_2 phase becomes crucial to understand the decomposition process for chlorapatite. However, as we mentioned above, our exhaustive examination with different experimental techniques did not uncover any CaCl_2 phase in either natural or synthetic sample. Interestingly, Murayama et al. (1986) synthesized tuite from hydroxylapatite [$\text{Ca}_3(\text{PO}_4)_2(\text{OH})$] and fluorapatite [$\text{Ca}_3(\text{PO}_4)_2\text{F}$] at high tempera-tures (1100–2300 °C) and pressures (10–15 GPa). The only phase they identified from the runs above 12 GPa is tuite, despite their intensive search for diffraction peaks ascribable to the possible decomposition components, such as Ca(OH)_2, CaF_2, or CaO. Murayama et al. (1986) assumed that this is probably because the fractions of possible decomposition components are too small (<10 wt%) and too widely dispersed in the polycrystalline texture of tuite to be detected by X-ray

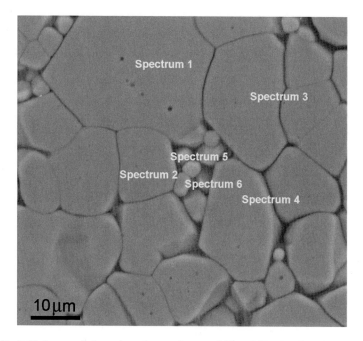

Fig. 5.31 BSE image of the enlarged central area of Fig. 5.27, showing the locations for energy-dispersive spectroscopic analyses

Table 5.10 EDS analyses of synthesized tuite crystals at six different locations (wt%)

Element	Sp*.1	Sp.2	Sp.3	Sp.4	Sp.5	Sp.6	Average
Ca	34.49	35.11	34.79	34.46	34.39	31.40	33.60
P	20.83	20.38	19.91	20.39	20.20	18.75	20.08
O	44.60	44.47	45.22	45.14	45.37	49.84	45.77
Cl	0.08	0.04	0.08	0.01	0.04	0.01	0.04
Total	100.00	100.00	100.00	100.00	100.00	100.00	

*Sp. spectrum

diffraction. We agree with Muyarama's explanation since we also found the complete or substantial loss of Cl in our synthetic or natural samples.

P-T Conditions

On the basis of the experimental data for chlorapatite (this study) and those for both hydroxylapatite and fluorapatite (Murayama et al. 1986), it appears that tuite can form at pressures as low as 10 GPa, and it is probably stable up to 20 GPa or higher (Murayama et al. 1986). This pressure stability range for tuite formed through the chlorapatite decomposition should be compared with that (up to 22 GPa) estimated from the other high-pressure phases in the veins of the Suizhou meteorite (Xie et al. 2002).

Geochemical and Mineralogical Significance

The volatile element-bearing apatites are known to be stable at shallower horizons of the upper mantle (Beswick and Carmichael 1978). At higher pressures and temperatures in the deep upper mantle, where volatile components are lacking, apatites eventually decompose to tuite (Murayama et al. 1986; Xie et al. 2003). Since the Earth's mantle has a chondritic composition with ~ 0.26 wt% P_2O_5 (Mason 1966), the behavior of phosphate minerals at depths in the terrestrial mantle where neither apatite nor whitlockite exists stably is of great interest for understanding the behavior of rare earth elements (REE) and other large lithophile elements. Consequently, tuite, stable at deeper mantle pressures and temperatures, could act as a potential host for REE, Na, Sr, and Ba (Murayama et al. 1986; Sugiyama and Tokonami 1987; Xie et al. 2003).

Chen et al. (1995a, b) reported that chlorapatite in shock veins of the Sixiangkou meteorite was transformed to an unknown high-pressure polymorph called phase A. This phase has a similar composition as the unshocked chlorapatite, but its Cl content (3.72 wt%) is lower than that for unshocked chlorapatite (4.48 wt%). Its Raman spectrum is also noticeably different from that for the unshocked hexagonal or monoclinic apatite, because the intense peak at 962 cm^{-1} for chlorapatite is split into three peaks at 960, 974, and 998 cm^{-1} in phase A, and the 430 and 591 cm^{-1} peaks are shifted to 414 and 582 cm^{-1}, respectively, but the 1081 cm^{-1} peak is shifted to 1092 cm^{-1} (Chen et al. 1995a, b). By comparing the Raman spectrum of phase A with that of our natural or synthetic tuite formed from the chlorapatite decomposition, we noticed some similarities between the Raman spectra of two phases, namely the intense peak at 974 cm^{-1} and the weak peak at 998 cm^{-1}; in addition, the three peaks at 414, 582, and 1092 cm^{-1} for phase A are similar to those for tuite. However, the 960 and 668 cm^{-1} peaks for phase A appear to match those for chlorapatite. In particular, the 960 cm^{-1} peak is still considerably strong. Accordingly, we believe that the Raman spectrum of phase A is actually a mixture of tuite and chlorapatite and the coexistence of tuite and chlorapatite implies an incomplete decomposition transformation of chlorapatite in the shock veins of the Sixiangkou meteorite. The rather high Cl content in phase A (3.72 wt%) bears additional evidence for our inference. Nevertheless, the nature of phase A in the Sixiangkou meteorite is still not fully understood, and the further investigation on this phase is needed.

5.2.8 Xieite, the $CaTi_2O_4$-Type Dense Polymorph of Chromite

Chromite is an important member of the spinel group. It is known that spinel-type AB_2O_4 compounds occur in many geological settings of the Earth's crust and mantle, as well as in lunar rocks and meteorites. High-pressure AB_2O_4 compounds are of great importance for the understanding of the constituents of the deep Earth.

More than forty years ago, in search of denser polymorphs of the then newly discovered silicate spinel (ringwoodite) and modified spinel (wadsleyite) that are stable at the pressure and temperature conditions of the Earth's transition zone, Ringwood proposed orthorhombic $CaFe_2O_4$-type and $CaTi_2O_4$-type structures as the top candidates for "post-spinel" transitions in the Earth's mantle (Reid and Ringwood 1969, 1970). High-pressure experiments revealed that AB_2O_4 compounds at high pressure might adopt $CaFe_2O_4$-, $CaMn_2O_4$- and $CaTi_2O_4$-type structures, whereby their structures attain atomic arrangements such that these structures become denser than spinel (Reid and Ringwood 1970). Experimental investigations have also indicated that the spinel compounds Mn_2O_3 and Fe_2O_3, and $MgFe_2O_4$ transform to a $CaFe_2O_4$-type structure at pressure above 25 GPa, that the $CaAl_2O_4$ and $MgAl_2O_4$ transform to a $CaFe_2O_4$ structure at pressure above 26.5 GPa, and that the $MgAl_2O_4$ structure changes to a $CaTi_2O_4$ structure at pressure above 40 GPa (Reid and Ringwood 1969; Mao et al. 1974; Irifine et al. 1991; Funamori et al. 1998; Akaogi et al. 1999; Fei et al. 1999; Andrault and Bolfan-Casanova 2001). In addition, $CaFe_2O_4$-type $NaAlSiO_4$ was experimentally shown to be stable at lower mantle conditions (Liu 1978; Irifune et al. 1994; Yagi et al. 1994; Tutti et al. 2000). However, no dense post-spinel polymorphs have been discovered in nature and confirmed as new minerals.

Chromite is one of the opaque accessory minerals in the Suizhou meteorite. During our recent study of the shock-related mineralogical features of the Suizhou meteorite, we discovered two post-spinel polymorphs of chromite (the dense $FeCr_2O_4$ phases) in the shock melt veins and in the areas directly adjacent to veins (Chen et al. 2003a, b). They are $CaTi_2O_4$- and $CaFe_2O_4$-structured $FeCr_2O_4$. Both these phases are new minerals found in nature.

In February 2008, the Commission on New Minerals, Nomenclature and Classification of the IMA approved this new mineral and its name "xieite" for the $CaTi_2O_4$-structured phase (IMA 2007-056). The mineral name, xieite, is from Prof. Xiande Xie, the former president of the IMA from 1990 to 1994, in honor of his contribution to mineralogy and shock effects of minerals.

The finding of above-mentioned two new minerals is important for understanding the structural characteristics of natural AB_2O_4 compounds under high pressure and temperature. The brief introductions of these two post-spinel dense polymorphs are sequentially described in this paragraph.

Occurrence

The chromite grains in the Suizhou meteorite usually contain abundant cracks and fractures (Figs. 3.43 and 3.44). However, the grains of chromite composition inside the Suizhou shock veins show smooth, unfractured surface and much brighter color in comparison with chromite outside the veins and with other high-pressure minerals in veins. This high-pressure phase has then been identified as the $CaTi_2O_4$-structured $FeCr_2O_4$, a dense post-spinel polymorph of chromite. It occurs as compact polycrystalline aggregates of about 5–40 μm in grain size and commonly displays as pseudomorph of chromite crystals or its fragments.

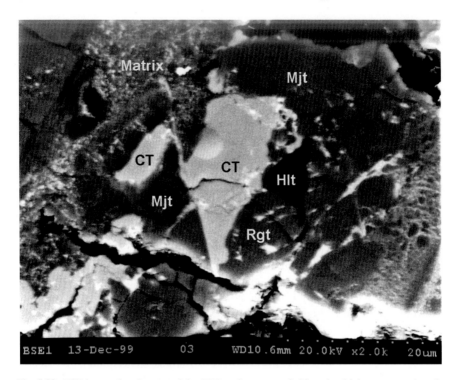

Fig. 5.32 BSE image showing two xieite (*CT*) grains surrounded by other high-pressure minerals. Rgt = ringwoodite, Mjt = majorite, Hlt = lingunite, Matrix = vein matrix

The grain sizes of xieite are relatively small (less than 20 μm). These small grains consist of dense phase only. Neither intergrowth of the dense phase and chromite, nor single grains of chromite have been observed inside the shock veins. Figure 5.32 is a BSE image displaying a semi-euhedral crystal of this dense phase that occurs in the close association with high-pressure minerals of the vein, such as ringwoodite, majorite, and lingunite. Figure 5.33 demonstrates a small single xieite grain of 10 μm × 4 μm in size embedded in the fine-grained vein matrix of the Suizhou meteorite. It is light gray in color and has smooth and unfractured surface.

Figures 3.34 and 5.35 show a xieite (CT) + chromite (Chr) intergrowth grain in a shock vein of the Suizhou meteorite. The xieite phase is directly contacting the vein wall, and the chromite phase is in the far side of the grain. It should be pointed out that there exists another high-pressure phase of chromite, the CF phase, in between the xieite and the chromite phases in this intergrowth grain.

Chemical Composition

The result of electron microprobe analysis of xieite is shown in Table 5.11. From this table, we can see that the chemical compositions of xieite are identical to those of chromite outside the veins (refer to Table 3.6). The main compositions of xieite are ~ 57 wt% Cr_2O_3 and 29 wt% FeO. The other minor components are Al_2O_3,

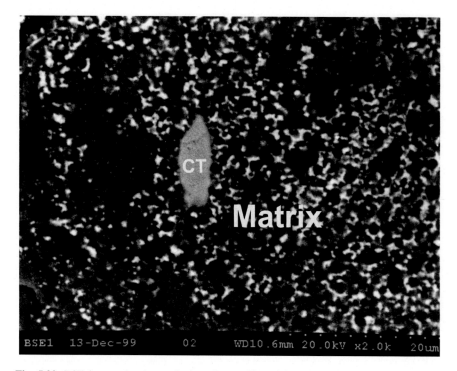

Fig. 5.33 BSE image showing a single polycrystalline xieite (*CT*) grain in the Suizhou vein matrix

MgO, TiO_2, MnO, and V_2O_3. The empirical formula of xieite is $(Fe_{0.87}Mg_{0.13}Mn_{0.01})_{1.01}-(Cr_{1.62}Al_{0.25}Ti_{0.08}V_{0.02})_{1.97}O_4$, based on 4 oxygen atoms per formula unit. The simplified formula for xieite is given as $FeCr_2O_4$.

The identity of chemical composition of the dense phase xieite and its host mineral chromite in their single grains indicates that this new phase formed directly from precursor chromite through solid-state phase transformation during a shock event. No additional elements were incorporated in the dense phase during its formation.

Physical and Optical Properties

Based on polishing relief, xieite is harder than chromite (a Mohs' hardness of 5.5). Calculated density is 5.63 g/cm^3 with empirical formula of xieite. Color of xieite under reflected light is gray. Bireflectance, pleochroism, anisotropy, and internal reflections could not be observed because of an occurrence of crystallite aggregate of xieite. The average reflectance values of xieite in air for the standard COM wavelengths are 19.9 % (470 nm), 19.7 % (546 nm), 18.6 % (589 nm), 17.6 % (65 nm), and 18.5 % (white light) (Chen et al. 2008).

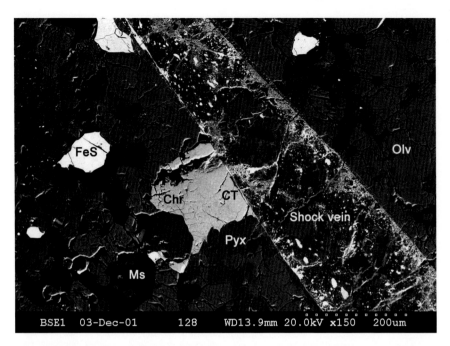

Fig. 5.34 BSE image showing a xieite (*CT*) + chromite (*Chr*) intergrowth grain contacting a shock vein in the Suizhou meteorite. Note that the CT phase is directly adjacent to the vein wall. Olv = olivine, Pyx = pyroxene, Ms = maskelynite, FeS = troilite

Raman Spectroscopy

There is a distinct difference in the feature of Raman spectra between xieite and chromite (Chen et al. 2008). The spectrum of xieite obtained by Raman microprobe shows a strong band at 605 cm^{-1} and a shoulder at 665 cm^{-1}, whereas the Raman spectrum of chromite displays distinct bands at 217, 280, 396, 595, and 680 cm^{-1} (Fig. 5.36). The Raman spectroscopic analyses demonstrate that the polycrystalline xieite grains within the shock veins consist of single phase of xieite, indicating a complete transformation from previous chromite to xieite (Figs. 5.30 and 5.31), whereas those chromite grains with lamellar texture in the chondritic portion adjacent to the shock vein show only partial transformation to xieite in the lamellar layers (Figs. 5.34 and 5.35). Chromite in chondritic portion far from the shock veins remains a spinel structure.

X-ray Diffraction Data

The crystal structure of xieite was measured in situ on polished sections of meteorite by using synchrotron X-ray diffraction analyses. Figure 5.37 shows two X-ray diffraction patterns recorded at different orientations of the sample, which show the polycrystalline nature of analyzed xieite grains. Table 5.12 lists the indexed peaks of X-ray diffraction pattern and Miller indices for xieite. The strongest lines in the X-ray powder diffraction pattern are [d(Å), I/I_o]: (2.675, 100), (2.389, 20), (2.089,

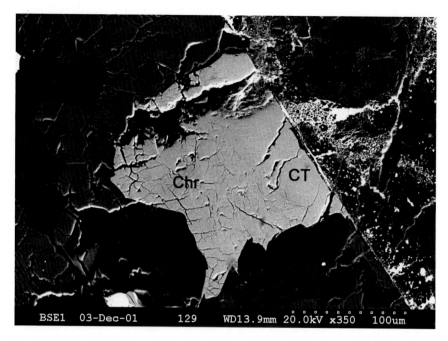

Fig. 5.35 Enlarged BSE image of the xieite (*CT*) + chromite (*Chr*) intergrowth grain contacting a shock vein. Note the smooth surface of xieite and the heavily fractured surface of chromite

Table 5.11 Compositions of xieite in/contacting the Suizhou shock veins (wt%)

Grain No.	1	2	3	4	5	Ave
MgO	2.63	2.74	2.60	2.59	2.53	2.62
FeO	29.61	29.51	29.74	29.67	29.98	29.70
MnO	0.82	0.83	0.73	0.79	0.88	0.81
CaO	n.d.	n.d.	n.d.	n.d.	n.d.	n.d.
TiO_2	2.61	2.33	2.68	2.59	2.73	2.59
Cr_2O_3	57.30	57.12	57.28	57.41	57.39	57.30
Al_2O_3	5.94	5.87	5.93	5.96	5.99	5.94
V_2O_3	0.97	0.91	0.96	1.02	0.99	0.97
Total	99.94	99.31	99.92	100.03	100.09	99.93

All data were measured by EPMA
1–4 single xieite grains in shock veins; *5* xieite in a three-zone grain contacting the vein; *n.d.* not detected

10), (1.953, 90), (1.566, 60), (1.439, 15), (1.425, 15), and (1.337, 40). Xieite is isostructural with synthesized $CaTi_2O_4$ with space group *Bbmm* and orthorhombic symmetry (Bertaut and Blum 1956; Bright et al. 1958). Cell parameters are as follows: a = 9.462 (6) Å, b = 9.562 (9) Å, c = 2.916 (1) Å (numbers in parentheses are standard deviations for the last significant digits), Z = 4, and V = 263.8 (4) Å^3.

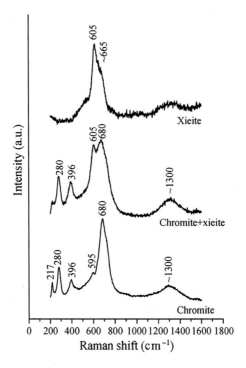

Fig. 5.36 Raman spectra of xieite (*upper*), its host chromite (*lower*), and a mixture consisting of both chromite and xieite (*middle*) (Chen et al. 2008)

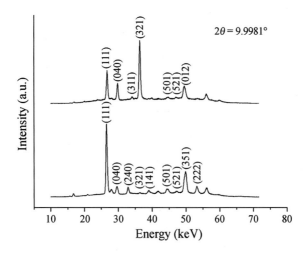

Fig. 5.37 X-ray diffraction patterns of polycrystalline aggregate of xieite obtained from two different orientations of the sample. The numbers listed at diffraction peaks are the Miller indices representative of the orientation of atomic planes in a crystal lattice (Chen et al. 2008)

Table 5.12 X-ray diffraction data of polycrystalline xieite[a]

h k l	d_{meas} (Å)	d_{cal} (Å)	I/I_0
1 1 1	2.6747	2.6754	100
0 4 0	2.3890	2.3905	20
2 4 0	2.1306	2.1336	5
4 2 0	2.1220	2.1203	5
3 1 1	2.0887	2.0895	10
3 2 1	1.9526	1.9542	90
1 4 1	1.8138	1.8144	8
5 0 1	1.5881	1.5875	10
5 1 1	1.5661	1.5661	60
5 2 1	1.5060	1.5066	5
0 1 2	1.4394	1.4414	15
3 5 1	1.4247	1.4263	15
2 2 2	1.3373	1.3377	40
4 1 2	1.2292	1.2309	2
6 0 2	1.0717	1.0706	6
2 6 2	1.0502	1.0490	5
6 3 2	1.0140	1.0149	8
6 4 2	0.9775	0.9771	7
1 2 3	0.9479	0.9477	2
3 2 3	0.9138	0.9119	3
3 3 3	0.8900	0.8918	4

[a]After Chen et al. (2008)

Fig. 5.38 Schematic view of xieite structure showing that Cr^{3+} and Al^{3+} occupy octahedral sites (BO_6), whereas Mg^{2+} and Fe^{2+} occupy dodecahedral sites (Chen et al. 2008)

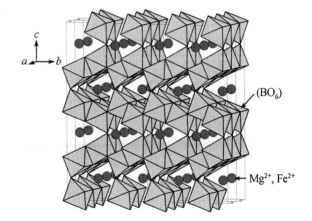

Figure 5.38 is a schematic view of xieite structure showing edge- and corner-sharing octahedral and dodecahedral sites, where cations of Cr^{3+} and Al^{3+} occupy octahedral sites, whereas Mg^{2+} and Fe^{2+} occupy dodecahedral sites.

Chen et al. (2003b) also synthesized the pure $CaTi_2O_4$-type phase at high pressures from 20 to 25 GPa and a temperature of 2000 °C, using laser-heated diamond anvil cells and a natural chromite crystal with a similar chemical composition as the chromite in the Suizhou meteorite as starting material. The quenched products recovered from the high P-T experiments were analyzed by synchrotron X-ray diffraction measurements and crystal structure determination. The X-ray diffraction pattern of the quenched $CaTi_2O_4$-type polymorph was indexed to give lattice parameters $a = 9.467$ (5) Å, $b = 9.550$ (7) Å, $c = 2.905$ (2) Å, $V = 262.6$ (4) Å^3, and $Z = 4$. The space group of this orthorhombic lattice is $Bbmm$. The d-spacings of the X-ray diffraction lines can be indexed in terms of the $CaTi_2O_4$ structure (Bright et al. 1958). The calculated density of the dense phase is 5.65 g/cm^3, which is 10.1 % denser than the original chromite (Chen et al. 2003b).

The above-obtained data suggest that both natural and synthetic xieite show almost identical cell parameters, thus giving confirmation of phase transformation of chromite in the Suizhou meteorite to the structure identical to $CaTi_2O_4$-structured phase synthesized experimentally.

Chromite–Xieite Transformation Mechanism

As we mentioned above, xieite occurs as compact polycrystalline aggregate grains and commonly displays as pseudomorph of chromite crystals or its fragments. Inside the shock veins, grains of chromite are completely transformed to massive xieite (Figs. 5.30 and 5.31). In the chondritic area adjacent to the shock veins, xieite may replace chromite completely for those grains closer to the vein, whereas those chromite grains relatively apart from the vein could only be replaced by lamellar xieite along fractures or special crystallography orientation (Figs. 5.32 and 5.33). Secondly, xieite has the same composition as chromite. Hence, we assume that the mineral xieite is formed through solid-state transformation of chromite, and its occurrence is intimately associated with the shock veins of meteorite.

P-T **Conditions**

The occurrence of xieite is related to a shock vein in the Suizhou meteorite, where peak shock pressure and temperature experienced by this meteorite were located. The P-T conditions for the formation of xieite can be constrained according to the high-pressure mineral assemblage within the shock vein. The occurrence of ringwoodite, majorite, lingunite, tuite, magnesiowüstite, and majorite–pyrope garnet in the shock vein constrains the peak pressure and temperature to 20–23 GPa and 1800–2000 °C, respectively (Agee et al. 1995; Chen et al. 1996a; Gillet et al. 2000; Xie et al. 2002). There must be a sharp gradient of temperature from the shock veins to the surrounding less shocked material, as the dense phase is not observed beyond 40 μm from the edge of a shock vein. It indicates that the high-pressure phase transitions are not kinetically obstructed in hot regions, especially inside and close to the shock veins. In comparison with experimentally unquenchable $CaMn_2O_4$-structured Fe_3O_4 (Fei et al. 1999) and $CaFe_2O_4$ (or $CaTi_2O_4$)-structured $ZnCr_2O_4$ polymorph (Wang et al. 2002), this natural xieite phase is quenchable during decompression.

Our experimental synthesis study shows that chromite transforms into xieite at 20–25 GPa and 2000 °C. Other experimental investigations indicate that, with increasing pressure, the $MgAl_2O_4$ spinel dissociates first to Al_2O_3 plus MgO above 25 GPa, and then, both oxides recombine to a $CaFe_2O_4$-type phase above 34 GPa and finally to a $CaTi_2O_4$-structured phase above 40 GPa (Funamori et al. 1998). We did not find any evidence indicating that chromite was decomposed into Cr_2O_3 plus FeO or recombined to a $CaTi_2O_4$-structured phase. Direct phase transition from $FeCr_2O_4$ to $CaTi_2O_4$-structure phase at pressure of 20–23 GPa and temperature of 1800–2000 °C is, therefore, inferred. According to the above high-pressure experiments, we assume that the $CaTi_2O_4$-structured phase may survive over an extended pressure range, from 20 up to 40 GPa, i.e., from the lower part of upper mantle down to the upper part of lower mantle.

Crystallochemical and Geochemical Significance

The $CaTi_2O_4$ structure is intimately related to the $CaFe_2O_4$ and $CaMn_2O_4$ structures. The orthorhombic $CaTi_2O_4$-, $CaFe_2O_4$-, and $CaMn_2O_4$-type structures are composed of dodecahedral (AO_8) and octahedral sites (BO_6), in which the differences among these structures lie in slight modifications of the polyhedral linkage (Reid and Ringwood 1970; Andrault and Bolfan-Casanova 2001). The only natural occurrence of minerals of any these structural types is marokite ($CaMn_2O_4$) that has been found in two terrestrial rocks formed at low pressure (Graudefroy et al. 1963; Villiers and Herbstein 1968). A $CaTi_2O_4$ crystal with space group *Bbmm* was synthesized experimentally (Bertaut and Blum 1956; Bright et al. 1958). No natural occurrence of $CaTi_2O_4$- and $CaFe_2O_4$-structured minerals has been reported prior to our finding of the dense $CaTi_2O_4$-type (xieite) and $CaFe_2O_4$-type phases of chromite in the Suizhou meteorite. In the spinel group (AB_2O_4), the cationic substitutions between Al^{3+} and Cr^{3+} and between Mg^{2+} and Fe^{2+} are extensive. Chrome-spinel $(Mg,Fe)(Al,Cr)_2O_4$ is an important accessory mineral in the Earth's mantle, as, for example, seen in lherzolite from the upper mantle (Scarfe et al. 1979). If the compound $(Mg,Fe)(Al,Cr)_2O_4$ or its analogues exist in the transition zone and the lower mantle, they might take the structure of $CaTi_2O_4$- or $CaFe_2O_4$- type compounds. Experiments have indeed demonstrated the existence of $CaTi_2O_4$- and $CaFe_2O_4$-structured $MgAl_2O_4$ phases at *P-T* conditions of the lower mantle (Irifine et al. 1991; Funamori et al. 1998). The natural $CaTi_2O_4$-structured $FeCr_2O_4$ phase (xieite) in the Suizhou meteorite contains about 6 wt% of Al_2O_3 and 2.6 wt% of MgO, whereby the cations Al^{3+} and Cr^{3+} occupy octahedral sites (BO_6) and the Mg^{2+} and Fe^{2+} occupy the dodecahedral sites (AO_8). Therefore, the natural occurrence of the $CaTi_2O_4$-type phase in the Suizhou meteorite indicates that the $CaTi_2O_4$ structure could be an important host phase for Cr^{3+} and Al^{3+}, and other metal elements (Mg^{2+}, Fe^{2+}, Ni^{2+}, Mn^{2+}, Zn^{2+}, and Mn^{3+}) in the deep Earth.

5.2.9 CF Phase, the CaFe₂O₄-Type Dense Polymorph of Chromite

As it was described in the previous paragraph, in the late 1960s, Ringwood proposed two orthorhombic $CaFe_2O_4$-type and $CaTi_2O_4$-type structures as the top candidates for "post-spinel" transitions in the Earth's mantle (Reid and Ringwood 1969, 1970), and besides xieite, the $CaTi_2O_4$-structured $FeCr_2O_4$ phase, we also found another post-spinel polymorph of chromite in the Suizhou meteorite. This second polymorph of chromite has just a $CaFe_2O_4$-type structure (Chen et al. 2003b). It occurs in the close association with the $CaTi_2O_4$-structured $FeCr_2O_4$ phase in the regions directly contacting the shock veins.

Occurrence
Petrographical studies in the Suizhou meteorite demonstrate the existence of a gradient of shock-produced pressure and temperature from the shock veins to the neighboring host meteorite during the shock event. Some chromite grains in close association with the shock veins covered a shock-produced pressure gradient. As it was mentioned above, xieite, the $CaTi_2O_4$-type polymorph of chromite, was found from these chromite grains, in addition to those occurring inside the shock veins (Chen et al. 2003b). In the electron back-scattering images (BSE) these chromite grains show three zones of distinct densities corresponding to the temperature gradient, i.e., xieite, the $CaTi_2O_4$-type phase, is close to the shock vein, chromite zone relatively apart from the vein, and a lamella-rich zone between xieite and chromite zones.

Figure 5.39 is a BSE image showing an intergrowth grain of chromite in the direct contact with a Suizhou shock vein. This grain is in the form of three zones consisting of the inner xieite phase, the intermediate lamella-like phase, and the outer chromite phase (Chen et al. 2003a, b). On that image, we can see that the xieite phase and the lamella-like phase are electron brighter than the chromite phase. Such an occurrence implies that the formation of the xieite phase and the lamella-like slices was highly pressure-dependent. The width of the lamellae in the lamella-like phase is ranging from 0.5 to 1.5 μm. Two to three sets of regular lamella-like slices in the lamella-like phase grain clearly indicate that the produce of these slices may intimately associate with special crystallographic orientation and shock-produced deformation features of parent chromite phase. Such a lamella-like zone between the xieite zone and chromite zone was firstly interpreted as a mixture of xieite and chromite (Chen et al. 2003a), but then identified as a new post-spinel phase of chromite that has the $CaFe_2O_4$-type structure (Chen et al. 2003b). It, thus, appears that this is a new second dense phase of chromite that was also formed due to shock transformation from precursor chromite.

Chemical Composition
Electron microprobe analyses of three zones in the intergrowth grain shown in Fig. 5.35 give a uniform chemical composition of chromite: 2.62 wt% MgO, 29.70 wt% FeO, 0.81 wt% MnO, 2.59 wt% TiO_2, 57.30 wt% Cr_2O_3, 5.97 wt%

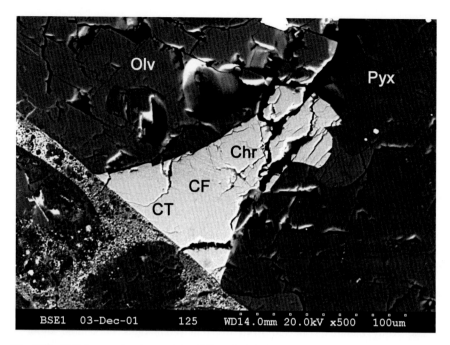

Fig. 5.39 BSE image showing a xieite (*CT*) + CF phase + chromite (*Chr*) intergrowth grain contacting a shock vein in the Suizhou meteorite. Note that the CF phase is in between the CT and chromite phases. Olv = olivine, Pyx = pyroxene (Chen et al. 2008)

Al_2O_3, 0.97 wt% V_2O_3, and 99.93 wt% in total. The chemical formula of the mineral chromite is $(Fe_{0.84}Mg_{0.14}Mn_{0.02})(Cr_{1.58}Al_{0.26}V_{0.03}Fe_{0.03}Ti^{4+}_{0.07})_{1.97}O_4$. The identity of chemical composition of this new dense $FeCr_2O_4$ phase, xieite and chromite, indicates that this new phase also formed directly from precursor chromite through solid-state phase transformation during a shock event. No additional elements were incorporated in the dense phase during its formation.

X-ray Diffraction Data

Our high *P-T* experiment indicates that the quenched CF and CT phases of chromite are difficult to distinguish with commonly used petrographic probes, such as petrographic microscopy, scanning electron microscopy, electron microprobe, and micro-Raman spectroscopy, but are clearly distinguishable by their characteristic X-ray diffraction patterns.

The X-ray diffraction patterns of both the natural and synthesized $CaFe_2O_4$-type phase of chromite were obtained using the same synchrotron radiation technique as we did for the xieite phase. For the natural sample, we focused the X-ray microprobe on the lamella-like zone between the xieite and chromite zones in an intergrowth chromite grain (Fig. 5.39) and obtained the diffraction pattern which is consistent with the $CaFe_2O_4$-type phase with a few lines of its neighboring phases, xieite and chromite (Fig. 5.40).

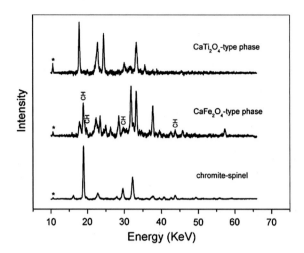

Fig. 5.40 X-ray diffraction patterns from chromite-spinel, CF, and CT phases. The peaks labeled with CH are from the residue of starting material chromite (Chen et al. 2003b). *Asterisk*, escape peaks

The X-ray patterns collected from different orientations of sample show a polycrystalline nature of the lamellae-like slices, with a preferential crystallographic orientation in the microcrystalline $CaFe_2O_4$-type phase (Fig. 5.41). We collected a total of 20 X-ray reflections from the natural $CaFe_2O_4$-type phase in addition to those from chromite (Table 5.13). They were indexed to an orthorhombic cell with

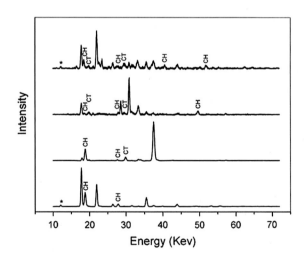

Fig. 5.41 X-ray diffraction patterns from natural CF phase. Each pattern was obtained at a different orientation of the sample. The peaks labeled with CH are from chromite and CT from CT phase. Other peaks that are unlabeled are from CF phase (Chen et al. 2003b). *Asterisk*, escape peaks

parameters $a = 8.954$ (7) Å, $b = 2.986$ (2) Å, $c = 9.891$ (7) Å, $V = 264.5$ (4) Å^3, and $Z = 4$ (Chen et al. 2003b). The space group of this orthorhombic lattice is *Pnma*. The calculated density of this natural dense phase is 5.62 g/cm^3, which is 9.4 % denser than the original chromite.

Chen et al. (2003b) synthesized a pure CaFe$_2$O$_4$-type phase with some residue of chromite above 12.5 GPa and 2000 °C. A total of 20 X-ray reflections from this phase were collected in addition to those from chromite (Table 5.13). The X-ray diffraction pattern of the quenched CaFe$_2$O$_4$-type polymorph was indexed to give lattice parameters $a = 8.955$ (7) Å, $b = 2.985$ (2) Å, $c = 9.909$ (7) Å, $V = 264.9$ (4) Å^3, and $Z = 4$. The space group of this orthorhombic lattice is *Pnma*. The calculated density of this natural dense phase is 5.61 g/cm^3.

It is evident that the crystal structure of our second natural dense phase of chromite is identical to that of the synthetic CaFe$_2$O$_4$-type phase obtained by high-pressure experiments. From the above data, we assume that the lamellae-like phase in the Suizhou meteorite is the first natural occurrence of CaFe$_2$O$_4$-type polymorph of chromite.

Table 5.13 Indexed peaks of the X-ray diffraction patterns and Miller indices collected from natural and synthetic CF phases (Chen et al. 2003b)

$h\ k\ l$	Natural		Synthetic	
	d_{obs} (Å)	d_{cal} (Å)	d_{obs} (Å)	d_{cal} (Å)
2 3 0	2.656 (2)$_s$	2.6549	2.657 (2)$_s$	2.6581
2 4 0	2.166 (2)$_s$	2.1647	2.169 (1)$_w$	2.1676
3 0 1	2.108 (3)$_w$	2.1111	2.108 (3)$_w$	2.1110
4 2 0	2.039 (1)$_s$	2.0394	2.035 (5)$_s$	2.0402
2 5 0	1.807 (2)$_w$	1.8095	1.808 (4)$_s$	1.8122
4 4 0	1.660 (3)$_s$	1.6596	1.663 (2)$_s$	1.6610
2 6 0			1.550 (7)$_w$	1.5498
5 0 1	1.536 (5)$_s$	1.5359		
0 0 2	1.498 (6)$_w$	1.4931	1.494 (2)$_s$	1 4928
5 2 1	1.468 (2)$_w$	1.4668	1.467 (1)$_w$	1.4670
1 6 1	1.425 (2)$_s$	1.4249	1.429 (2)$_s$	1 4267
2 6 1	1.372 (1)$_w$	1.3737	1.370 (5)$_w$	1.3753
3 0 2	1.336 (1)$_w$	1.3354	1.334 (1)$_w$	1.3352
3 2 2	1 285 (4)$_w$	1.2892	1.287 (2)$_w$	1.2892
1 4 2	1.267 (2)$_s$	1.2654	1.263 (2)$_s$	1.2658
4 2 2	1.206 (2)$_w$	1.2047	1.206 (2)$_w$	1.2048
5 2 2	1117 (1)$_w$	1.1172	1.115 (2)$_w$	1.1172
5 3 2	1.081 (2)$_w$	1.0832	1.084 (1)$_w$	1.0833
4 2 3	0.893 (1)$_w$	0.8945	0.891 (3)$_w$	0.8945
1 6 3	0.848 (1)$_w$	0.8483	0.849 (1)$_w$	0.8486
6 0 3	0.828 (2)$_w$	0.8281	0.829 (1)$_w$	0.8280

d_{obs} and d_{cal} are observed and calculated d-values, respectively; $_s$, strong diffraction peak; and $_w$, weak diffraction peak

Chromite–CF Phase Transformation Mechanism

The CF phase occurs only in the form of lamella-like zone between the CT and chromite-spinel zones in shocked chromite grains. No individual grains of CF phase were observed so far in the Suizhou meteorite. Furthermore, the CF phase has the same composition as its precursor chromite. So, the occurrence and mineral composition of this CF-structured polymorph of chromite in the Suizhou meteorite support a solid-state mechanism for the transformation of chromite to CF phase. There is no evidence for an intermediate decomposition process, such as the post-spinel transitions of $MgAl_2O_4$ (Funamori et al. 1998). The results indicate that the solid-state phase transformation is, in general, a dominating mechanism for the formation of high-pressure minerals in naturally shocked meteorites (Chen et al. 1996a).

***P-T* Conditions**

The occurrence of the $CaTi_2O_4$- and $CaFe_2O_4$-structured phases of chromite is related to a shock vein in the Suizhou meteorite, where peak shock pressure and temperature experienced by this meteorite were located. The *P-T* conditions for the formation of both $CaTi_2O_4$- and $CaFe_2O_4$-structured phases can be constrained according to the high-pressure mineral assemblage within the shock vein. The occurrence of ringwoodite, majorite, $NaAlSi_3O_8$ hollandite, tuite, and majorite–pyrope garnet in the shock vein constrains the peak pressure and temperature to 20–23 GPa and 1800–2000 °C, respectively (Agee et al. 1995; Chen et al. 1996a; Gillet et al. 2000; Xie et al. 2002).

The presence of a sharp gradient of temperature from the shock veins to the surrounding less shocked material, the $CaTi_2O_4$- and $CaFe_2O_4$-structured phases, as two dense phases of chromite should be experienced high-pressure and high-temperature regime close or a bit lower than the *P-T* conditions in the veins themselves. The pressure of 18–23 GPa and temperature of 1800–2000 °C are estimated for the shocked intergrowth grains of chromite, for which a solid-state transformation from chromite to CF phase takes place at about 18–23 GPa and ≤1800 °C.

Crystallochemical and Geochemical Significance

It is well known that the chromite $(Mg,Fe)(Al,Cr)_2O_4$ is an important accessory mineral in the Earth's mantle. If the compound $(Mg,Fe)(Al,Cr)_2O_4$ or its analogues exist in the transition zone and the lower mantle, they might take the structure of $CaTi_2O_4$- or $CaFe_2O_4$-type compounds. Experiments have indeed demonstrated the existence of $CaTi_2O_4$- and $CaFe_2O_4$-structured $MgAl_2O_4$ phases at *P-T* conditions of the lower mantle (Irifine et al. 1991; Funamori et al. 1998). The natural $CaFe_2O_4$-structured $FeCr_2O_4$ phase in the Suizhou meteorite contains about 6 wt% of Al_2O_3 and 2.6 wt% of MgO, whereby the cations Al^{3+} and Cr^{3+} occupy octahedral sites (BO_6) and the Mg^{2+} and Fe^{2+} occupy the dodecahedral sites (AO_8). Therefore, the natural occurrence of the $CaFe_2O_4$-structured $FeCr_2O_4$ phase in the Suizhou meteorite indicates that besides the $CaTi_2O_4$ structure, the $CaFe_2O_4$ structure could also be an important host phase for Cr^{3+}, Al^{3+}, and other metal elements (Mg^{2+}, Fe^{2+}, Ni^{2+}, Mn^{2+}, Zn^{2+} and Mn^{3+}) in the deep Earth (Chen et al. 2003b).

5.2.10 Multi-phase Grains of Silicate and Oxide Minerals

It is revealed that some unusual occurrences of shock-produced high-pressure phases were observed in the Suizhou shock veins. They are two-phase or three-phase grains. These grains or fragments consist of high-pressure polymorphs of two or three silicate minerals, namely ringwoodite, majorite, and lingunite (Figs. 5.42, 5.43, and 5.44). The boundaries between different high-pressure polymorphs of silicate minerals are sharp and even, or slightly curved, and sometimes irregular. Besides, Xie et al. (2011b) reported rather distinct two-phase grains consisting of xieite and a high-pressure polymorph of one of the above-mentioned three silicate minerals (Fig. 5.45). These distinct two-phase grains are small, 20–25 μm in length, and show rounded or drop-like outlines.

It is interesting that the boundaries between high-pressure polymorphs of silicate or oxide minerals are quite sharp (Figs. 5.44 and 5.45), implying that partial melting

Fig. 5.42 BSE image showing a two-high-pressure-phase grain of ringwoodite (*Rgt*) and lingunite (*Hlt*) in the Suizhou vein matrix. Note the clear boundaries between these two phases

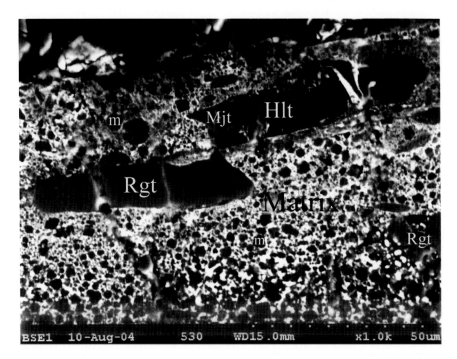

Fig. 5.43 BSE image showing a two-high-pressure-phase grain of majorite (*Mjt*) and lingunite (*Hlt*) in the Suizhou vein matrix (*Matrix*). Rgt = ringwoodite, m = majorite–pyrope garnet

does not take place at the boundary areas. All these two-phase and three-phase grains show smooth surface, and no zoning and no fractures or cracks were observed. All mineral phases in these multi-phase grains were identified by both Raman microprobe spectroscopy (Fig. 5.46) and chemical analysis. The Raman spectrum of xieite in a two-phase grain gives Raman bands at 536, 607 and 666 cm^{-1} (Fig. 5.46a) which is identical to that of xieite in its single grain (Fig. 5.36). The Raman spectra of the three silicate high-pressure minerals in two-phase grains (Fig. 5.46b–d) are also similar to those in their single grains (Figs. 5.5a, 5.14a and 5.16$_1$). In this section, we shall lay our stress on the two-phase grains of xieite and a silicate mineral.

5.2.10.1 Ringwoodite + Lingunite Grain

The two-high-pressure-phase grains of ringwoodite + lingunite were observed in shock veins of the Suizhou meteorite. Figures 5.42 and 5.47 are BSE images showing two such grains in the vein matrix. These two grains are of elongated elliptic shape with 80 and 110 μm in length and 40 and 50 μm in width, respectively. In these two grains, ringwoodite is the major high-pressure phase and lingunite takes up secondary position. However, ringwoodite in this grain was fractured (Fig. 5.42) or broken (Fig. 5.47), but lingunite shows intact feature.

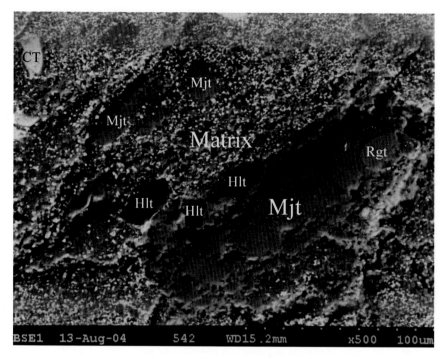

Fig. 5.44 BSE image showing a large three-phase grain of ringwoodite (*Rgt*), majorite (*Mjt*), and lingunite (*Hlt*) in the Suizhou shock vein matrix (*Matrix*). CT = xieite

Under SEM in back-scattered mode, the boundaries between the gray-colored ringwoodite and the black-colored lingunite are quite sharp, which may reflect the original occurrence of their precursor minerals olivine and plagioclase in these two meteorite fragments. No trace of partial melting in boundary areas was observed in both grains.

5.2.10.2 Lingunite + Majorite Grains

The two-high-pressure-phase grains of lingunite + majorite were also observed in shock veins of the Suizhou meteorite. Figure 5.43 is a BSE image showing such a two-high-pressure-phase grain of lingunite and majorite in the vein matrix. This two-phase grain has elongated shape with 60 μm in length and 20 μm in width. In this grain, lingunite is the main high-pressure phase and majorite takes secondary position. The surfaces of both lingunite and majorite are smooth, and no fractures or cracks were observed in these two high-pressure minerals.

Under SEM in back-scattered mode, the boundaries between the black-colored lingunite and the dark gray-colored lingunite are also quite sharp, which may reflect the original occurrence of their precursor minerals such as plagioclase and low-Ca

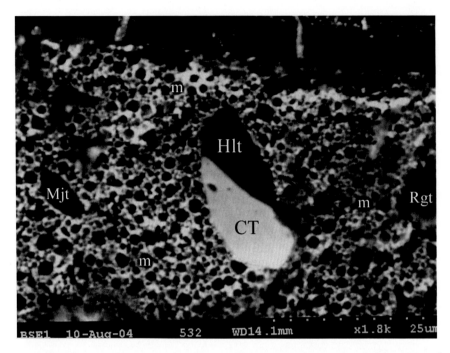

Fig. 5.45 BSE image showing a drop-like two-high-pressure-phase grain of xieite (*CT*) and lingunite (*Hlt*) in the Suizhou vein matrix. Rgt = ringwoodite, Mjt = majorite, m = majorite–pyrope garnet (Xie et al. 2011a)

pyroxene in that meteorite fragment. No trace of partial melting in boundary area between these two phases was detected.

5.2.10.3   Ringwoodite + Majorite + Lingunite Grains

The three-high-pressure-phase grains of ringwoodite + majorite + lingunite were also observed in shock veins of the Suizhou meteorite. Figure 5.44 is a BSE image showing a three-high-pressure-phase grain of ringwoodite, majorite, and lingunite in the shock vein matrix. This three-phase grain also has elongated oval shape with 150 μm in length and 80 μm in width. In this large grain, majorite is the main high-pressure phase. There are only one small ringwoodite and two small lingunite fragments embedded in majorite phase. The surfaces of both ringwoodite and lingunite fragments are smooth, and no fractures or cracks were observed in these two minerals.

Under SEM in back-scattered mode, the boundary between the gray-colored ringwoodite and the dark gray-colored majorite and the boundaries between the black-colored lingunite and the dark gray-colored majorite are also quite sharp, which may reflect the original occurrence of their precursor minerals such as olivine, low-Ca pyroxene, and plagioclase in that meteorite fragment. No trace of partial melting in boundary area between these three-high-pressure phases was observed.

Fig. 5.46 Raman spectra of minerals in two-phase grains. **a** Xieite, **b** lingunite, **c** ringwoodite, and **d** majorite (Xie et al. 2011a)

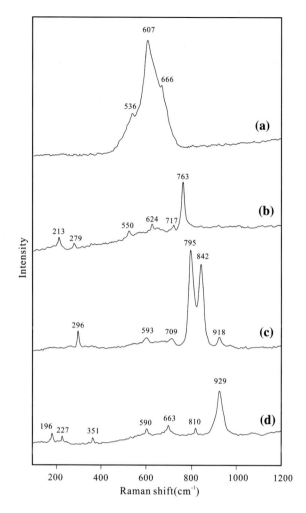

5.2.10.4 Ringwoodite + Majorite + Perovskite Glass Grains

The three-phase grains of ringwoodite + majorite + perovskite glass were also observed in shock veins of the Suizhou meteorite (Chen et al. 2004c). Figure 5.48 is a BSE image showing an ovoid-shaped three-phase grain of ringwoodite, majorite, and perovskite glass in the vein matrix. This three-phase grain is of 100 μm in length and 40 μm in width. This grain is little bit different with other three-high-pressure-phase grains because it, originally, was a two-phase fragment consisted of pyroxene and olivine. At the shock-produced high pressure and high temperature, olivine transformed to ringwoodite, but pyroxene transformed to its two high-pressure polymorphs, namely perovskite in the interior area due to a relatively low

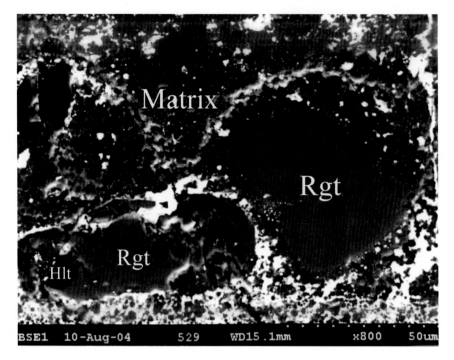

Fig. 5.47 BSE image showing a two-phase grain of ringwoodite (*Rgt*) and lingunite (*Hlt*) in the Suizhou shock vein matrix (*Matrix*)

temperature and majorite at the rim region due to a higher temperature. After pressure release, perovskite vitrified at post-shock temperature (Chen et al. 2004c).

In this large grain, the surfaces of both perovskite glass and ringwoodite are smooth and even, and no fractures or cracks were observed in perovskite glass interior, and only one fine fracture was seen in ringwoodite. However, quite a lot irregular cracks developed in majorite rim because of the volume increase induced by the vitrification of perovskite.

Under SEM, the boundary between perovskite glass interior and majorite rim is rough and uneven. Microprobe analyses show that the low-Ca pyroxene in the Suizhou chondritic host, the majorite, and the perovskite glass phase in the ovoid grains are identical in composition (Chen et al. 2004c). This indicates that no element exchange between these three phases took place during shock metamorphism.

5.2.10.5   Xieite + Lingunite Grains

An unusual occurrence of shock-produced high-pressure phases in the Suizhou melt veins is the two-phase grains consisting of xieite and a high-pressure polymorph of

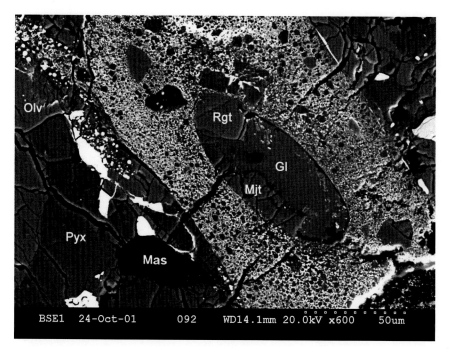

Fig. 5.48 BSE image showing an ovoid-shaped three-phase grain of ringwoodite (*Rgt*), majorite (*Mjt*), and perovskite glass (*Gl*) in a Suizhou shock vein. Olv = Olivine, Pyx = pyroxene, Mas = maskelynite

one of the three main rock-forming silicate minerals, namely lingunite, ringwoodite, and majorite (Xie et al. 2011b). However, the rather common two-phase grains are consisted of xieite + lingunite. Figure 5.45 is a BSE image showing such a two-phase grain in the Suizhou vein matrix. It is interesting that this grain has a drop-like shape and its size is relatively small (30 μm × 13 μm). The volumes of both xieite and lingunite in this grain are almost equal, and the boundary between these two phases is sharp, but little bit curved. It implies that no obvious partial melting in boundary area took place. Figure 5.49 shows another type of xieite and lingunite two-phase grain of larger size (80 μm × 25 μm), where xieite is the main body, and lingunite take up only very small portion of this grain. The boundary between two phases is also sharp but curved, implying that both xieite and lingunite might experience partial melting in the boundary area.

Compositions of xieite and lingunite in the two-phase grain were measured by EDS technique, and the results are shown in Table 5.14. For comparison, the compositions of their precursors such as chromite and plagioclase outside the Suizhou shock vein were also measured and shown in this table. It is revealed that the xieite contains higher content of Al_2O_3 (6.19 vs. 5.45 wt%) and lower content of Cr_2O_3 (56.48 vs. 57.98 wt%) than its precursor chromite. On the other hand, the lingunite contains higher contents of Cr_2O_3 (0.32 vs. 0.03 wt%) and FeO (1.97 vs.

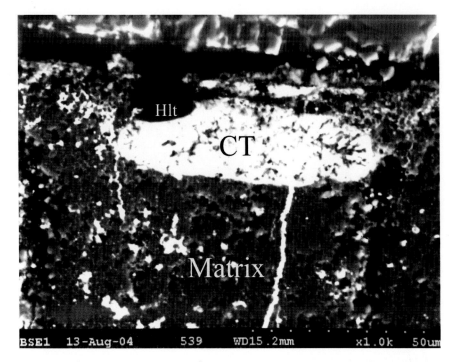

Fig. 5.49 BSE image showing a two-phase grain of xieite (*CT*) + lingunite (*Hlt*) in the Suizhou shock vein matrix (*Matrix*). Note the curved boundary between xieite and lingunite phases

0.41 wt%) and lower contents of Al_2O_3 (19.98 vs. 21.73 wt%) than its precursor plagioclase (Xie et al. 2011b). This implies that the element diffusion appeared between the two phases in which some of the chromium and iron migrated from xieite to lingunite and some of the aluminum migrated from lingunite to xieite.

5.2.10.6 Xieite + Ringwoodite Grains

The two-high-pressure-phase grains of xieite + ringwoodite were also observed in shock veins of the Suizhou meteorite (Xie et al. 2011b). Figure 5.50 is a BSE image showing such a two-phase grain in the vein matrix. Its grain size is 30 μm long and 15 μm wide. This grain originally has an ovoid form, but its lower and left parts were ground off during thin section preparation. The volumes of both xieite and lingunite phases in this grain are almost equal, and the boundary between these two phases is sharp, but slightly curved. This implies that no obvious partial melting in boundary area took place.

Compositions of xieite and ringwoodite in the two-phase grain were also measured by EDS technique, and the results are shown in Table 5.15. For comparison, the compositions of their precursors such as chromite and olivine outside the

Table 5.14 Composition of xieite and lingunite in two-phase grain and their precursors (wt%)

	Chromite in chondrite	Xieite in two-phase grain	Lingunite in two-phase grain	Plagioclase in chondrite
SiO$_2$			65.8	65.57
Al$_2$O$_3$	5.5	6.2↑	20.0↓	21.73
MgO	2.3	2.4		0.00
FeO	30.4	30.8	2.0↑	0.41
CaO			2.3	2.12
MnO	–	0.7	–	0.02
TiO$_2$	2.9	2.4		0.04
Cr$_2$O$_3$	58.0	56.5↓	0.3↑	0.03
V$_2$O$_3$	0.9	1.0		
Na$_2$O			8.4	8.87
K$_2$O			1.2	1.31
Total	100.0	100.0	100.0	100.38

Plagioclase in Suizhou chondrite was analyzed by EPMA. All other data were measured by EDS technique

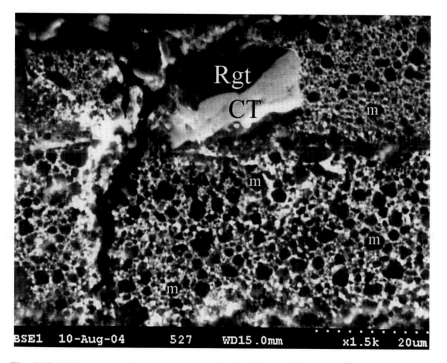

Fig. 5.50 BSE image showing a two-phase grain of xieite (*CT*) + ringwoodite (*Rgt*) in the Suizhou vein matrix (*Matrix*). Note the curved boundary between xieite and ringwoodite phases. M = majorite–pyrope garnet (Xie et al. 2011a)

Table 5.15 Composition of xieite + ringwoodite in two-phase grain and their precursors (wt%)

	Chromite in chondrite	Xieite in two-phase grain	Ringwoodite in two-phase grain	Olivine in chondrite
SiO_2			39.9	38.3
Al_2O_3	5.5	5.9		
MgO	2.3	2.2	39.5	39.4
FeO	30.4	32.0↑	20.0↓	22.3
MnO	–	0.7		
TiO_2	2.9	2.5		
Cr_2O_3	58.0	55.6↓	0.6↑	0.0
V_2O_5	0.9	1.1		
Total	100.0	100.0	100.0	100.0

All data were measured by EDS technique

Suizhou shock vein were also measured and shown in this table. It is revealed that the general phenomena of element diffusion in xieite + ringwoodite grains are similar to that in xieite + lingunite grains, but the Al_2O_3 constituent is replaced by FeO, since olivine does not contain Al_2O_3. Comparing to the precursor mineral chromite, the xieite contains higher content of FeO (31.96 vs. 30.39 wt%) and lower content of Cr_2O_3 (55.57 vs. 57.98 wt%). On the other hand, in comparison with the precursor mineral olivine, the content of FeO in ringwoodite decreased from 22.35 to 19.95 wt% and the content of Cr_2O_3 increased from 0.03 to 0.58 wt% (Xie et al. 2011b). This implies that the element diffusion also took place between these two phases in which some of the chromium migrated from xieite to ringwoodite and some of the iron migrated from ringwoodite to xieite.

5.2.10.7 Xieite + Majorite Grains

The two-high-pressure-phase grains of xieite + majorite can also be observed in shock veins of the Suizhou meteorite. Figure 5.51 is a BSE image showing such a two-phase grain in the vein matrix (Xie et al. 2011b). Its grain size is only 20 μm long and 18 μm wide. Xieite phase in this grain has an ovoid form, but majorite phase shows irregular form. The boundary between these two phases is sharp enough and curved. This implies that no obvious partial melting in boundary area took place.

Compositions of xieite and majorite in the two-phase grain were measured by EDS technique, and the results are shown in Table 5.16. For comparison, the compositions of their precursors such as chromite and low-Ca pyroxene outside the Suizhou shock vein were also measured and shown in this table. From this table, we can see that the situation of element diffusion in xieite + majorite grains is remarkably different from that for the above-described two cases. On the whole, the composition of xieite in this two-phase grain is similar to that of its precursor chromite, only with slight increase in Al_2O_3 (5.69 vs. 5.45 wt%) and very little

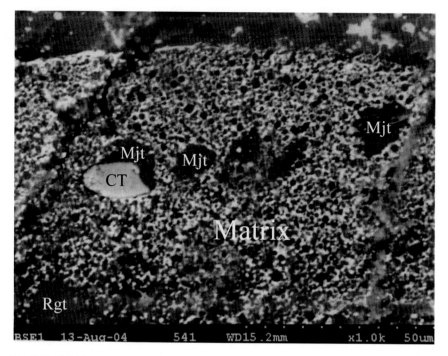

Fig. 5.51 BSE image showing a two-phase grain of xieite (*CT*) + majorite (*Mjt*) in the Suizhou vein matrix (*Matrix*). Note the curved boundary between xieite and majorite phases. Rgt = ringwoodite (Xie et al. 2011a)

Table 5.16 Composition of xieite + majorite in two-phase grain and their precursors (wt%)

	Chromite in chondrite	Xieite in two-phase grain	Majorite in two-phase grain	Pyroxene in chondrite
SiO$_2$			50.7↓	55.8
Al$_2$O$_3$	5.5	5.7	4.3↑	0.2
MgO	2.3	2.0↓	27.6↓	29.3
FeO	30.4	31.3↑	14.8↑	13.9
CaO			1.7 ↑	0.7
MnO	–	0.7		
TiO$_2$	2.9	2.7		
Cr$_2$O$_3$	58.0	57.6↓	0.9↑	0.1
V$_2$O$_3$	0.9	1.0		
Total	100.0	100.0	100.0	100.0

All data were measured by EDS technique

decrease in Cr_2O_3 (57.54 vs. 57.98 wt%), but the content of FeO increased from 30.39 to 31.32 wt%, and the content of MgO decreased from 2.32 to 2.00 wt%. On the other hand, the chemical composition of majorite in this two-phase grain is significantly different from that of its precursor low-Ca pyroxene. The main difference is shown in that the majorite shows remarkable decrease of SiO_2 (50.71 vs. 55.78 wt%) and MgO (27.61 vs. 29.31 wt%); remarkable increase of Al_2O_3 (4.29 vs. 0.16 wt%); and small increase of FeO (14.76 vs. 13.95 wt%), CaO (1.73 vs. 0.70 wt%), and Cr_2O_3 (0.91 vs. 0.10 wt%) (Xie et al. 2011b). This indicates that the element exchange between xieite and majorite in this kind of two-phase grain is relatively limited. Only very small amount of aluminum and iron migrated from majorite to xieite, and very little amount of chromium migrated from xieite to majorite. However, the remarkable difference in composition between majorite and its precursor pyroxene demonstrates that some majorite components, such as Al_2O_3 and CaO, must be captured from the surrounding shock-induced silicate melt. Hence, there exist two types of element exchange for xieite + majorite grains: exchange between two phases inside the grain and exchange between majorite inside the grain and Al- and Ca-bearing silicate melt outside the grain.

5.3 Fine-Grained High-Pressure Minerals in Shock Veins

It is known that the main constituent of shock veins in Suizhou is fine-grained matrix that makes up 80–90 % of the veins by volume. Our SEM, EPMA, and Raman spectroscopic investigations showed that the matrix material consists mainly of isometric granular garnet of majorite–pyrope composition, irregular FeNi metal, and FeS grains in eutectic intergrowths which filled the interstices of garnet grains (Fig. 5.52). Their grain sizes are ranging from 0.2 to 2 μm (Xie et al. 2001a).

In order to identify the extremely small-sized vein matrix minerals and to reveal their crystallinity, 5 areas on a polished thin section were selected for synchrotron radiation X-ray diffraction in situ analysis (Xie et al. 2005b). The X-ray beam is 15 μm × 15 μm in size that covers an area of 225 μm² on the polished thin section. This would give us good X-ray powder diffraction patterns of matrix minerals since the volume of each of the X-ray beamed sample will be of 6750 μm³ (225 μm² × 30 μm, the later is the thickness of probed thin section), and such a volume would contain at least 6750 fine grains of matrix minerals if we take 1 μm in diameter as the mean grain size for all matrix minerals. In fact, we did obtain beautiful patterns, which clearly show sharp diffraction peaks of all constituent minerals in vein matrix, and the resolution of diffraction lines was high enough that almost no overlapping of diffraction peaks was observed (Fig. 5.53). Following results were obtained from the X-ray diffraction analyses:

(i) The vein matrix is really consisted of three minerals, namely majorite garnet (the high-pressure polymorph of pyroxene), kamacite (FeNi metal), and troilite (FeS) (Fig. 5.53). The characteristic diffraction lines of these minerals are as

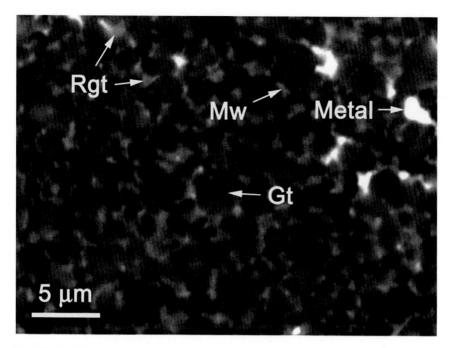

Fig. 5.52 BSE image showing the fine-grained high-pressure assemblage consisting of majorite–pyropess garnet (*Gt*), magnesiowüstite (*Mw*), and ringwoodite (*Rgt*). Metal = FeNi metal

 follows: majorite garnet: d = 2.571, 2.88, 2.451, 1.538, 2.349, 2.255, and 1.596 Å; FeNi metal: d = 2.026, 1.170, and 1.453 Å; and troilite: d = 2.091, 2.667, 2.99, 1.934, and 1.728 Å.

(ii) The vein matrix is rather homogeneous in mineral composition, e.g., all three constituent minerals can be detected by every analysis. However, each of the minerals may not be so evenly distributed in vein matrix. Figure 5.53a shows that the probed area is consisted of majorite–pyrope garnet, kamacite, and troilite, but the diffraction peaks of eutectic intergrowths of kamacite and troilite are stronger than those of majorite–pyrope garnet. Figure 5.53b demonstrates that the probed area is also consisted of majorite–pyrope garnet, kamacite, and troilite, but garnet and kamacite show similar intensities.

(iii) All the X-ray powder diffraction patterns we obtained show sharp or rather sharp diffraction peaks with good resolution for all three constituent minerals, and no amorphous phases in any of the probed areas were found, indicating good crystallinity of the minerals. This implies that all the shock-induced molten materials in veins had long enough time (a few seconds) to crystallize into different minerals with good crystallinity under pressure (Chen et al. 1996a), and the cooling rate of the veins was big enough to survive all crystalline phases after pressure release.

Fig. 5.53 Synchrotron
radiation X-ray diffraction
patterns of fine-grained vein
matrix in the Suizhou
meteorite. Patterns **a** and
b obtained at two different
probed areas on a polished
section. Mj = majorite–pyrope
garnet in solid solution,
FeNi = FeNi metal,
Tr = troilite (Xie et al. 2005a)

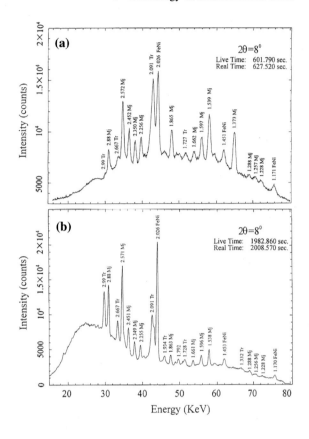

5.3.1 Majorite–Pyrope Garnet

One of the shock-induced effects on the Suizhou meteorite is the melting of silicate
minerals, such as olivine, pyroxene, and plagioclase, in veins. The shock-produced
silicate melt is a liquid phase of SiO_2, $(Mg,Fe)O$, Al_2O_3, Na_2O, and CaO. The main
mineral phase crystallized from this melt under pressure is majorite–pyrope garnet.

Occurrence
Majorite–pyrope garnet occurs as tiny idiomorphic crystals ranging from 0.3 to
2.5 μm in diameter in the vein matrix of Suizhou (Fig. 4.6). Under SEM in
back-scattered mode, garnet is black in color and has smooth surface. No fractures
were observed in garnet crystals. Generally, majorite–pyrope garnet crystals are
surrounded by magnesiowüstite and microcrystalline ringwoodite, but sometimes
by FeNi metal + FeS intergrowth grains.

Chemical Composition
The electron microprobe analyses show that in contrast to the coarse-grained
polycrystalline low-Ca majorite, these idiomorphic matrix garnets are richer in

Table 5.17 Chemical composition of majorite–pyrope$_{SS}$ in the Suizhou shock veins (wt%)

	SZ-MjPr-1	SZ-MjPr-2	SZ-MjPr-3	Mjt-Pyr$_{SS}$ average	Majorite average	Pyroxene average
SiO$_2$	50.279	50.209	51.094	50.527↓	55.797	55.775
TiO$_2$	0.135	0.075	0.149	0.120	0.191	0.162
Al$_2$O$_3$	3.605	3.672	3.260	**3.512**↑	0.123	0.158
Cr$_2$O$_3$	0.372	0.447	0.169	**0.329**↑	0.121	0.110
MgO	27.866	28.004	28.518	28.129	28.891	29.239
CaO	1.976	1.835	1.943	**2.918**↑	0.710	0.715
MnO	0.371	0.353	0.368	0.364	0.499	0.496
FeO	13.777	13.739	13.309	13.608↓	14.365	13.952
NiO	0.197	0.232	0.301	**0.243**↑	0.076	0.007
Na$_2$O	0.927	0.842	0.840	**0.870**↑	0.119	0.043
K$_2$O	0.033	0.048	0.057	**0.046**↑	0.023	0.009
Total	99.534	99.455	100.188	100.666	100.915	100.666
Fs	6.25	6.39	6.02	6.21	20.85	20.12
En	89.20	89.40	89.77	89.45	77.78	78.41
Wo	4.55	4.21	4.31	4.34	1.37	1.38

All values were determined by EPMA in wt%
Fs—Ferrosite, En—Enstatite, Wo—Wollastonite

Al$_2$O$_3$, CaO, Na$_2$O, and K$_2$O and contain appreciable amount of Cr$_2$O$_3$ and NiO (Table 5.17). The rather high content of Al$_2$O$_3$ (up to 3.51 wt%) indicates that the garnet is constituted not only of majorite composition, but also of pyrope composition. Hence, we call this liquidus garnet phase as majorite–pyrope solid solution. Since the main silicate minerals such as olivine and low-Ca pyroxene in the Suizhou unmelted chondritic rock contain little Al, Na, and Cr, the majorite–pyrope solid solution could not have formed by direct transformation from these phases, but rather crystallized from a shock-induced melt that was enriched in Al$_2$O$_3$, CaO, Na$_2$O, K$_2$O, and Cr$_2$O$_3$. Such phenomenon was firstly observed in the Sixiangkou shock veins, where the majorite–pyrope solid solution scavenged all of the Na, a majority of the Al and Ca, a part of the Cr in the shock-induced Sixiangkou melt (Chen et al. 1996a).

Raman Spectroscopy

The fine-grained garnet of majorite–pyrope composition shows Raman spectrum with peaks at 928, 664, 536, and 350 cm^{-1} (Fig. 5.7b) which is quite different with the Raman peaks of pyroxene outside the veins (Fig. 3.7c). Although the Raman pattern of fine-grained majorite–pyrope garnet is similar to that of the coarse-grained majorite, the intensity of the Raman peak at 664 cm^{-1} is higher than that of the coarse-grained majorite (Fig. 5.7a). This kind of Raman spectra is characteristic for liquidus majorite garnet.

Synchrotron Radiation X-ray Diffraction Data

The majorite–pyrope grains in the Suizhou shock veins are relatively small (0.3–2.5 μm in diameter) and they are in close association with fine-grained FeNi metal and troilite, so it is difficult to obtain single-phase sample of this mineral for X-ray diffraction analysis. However, we succeeded in getting its X-ray diffraction pattern using the in situ synchrotron radiation microprobe technique.

As it was introduced in Chap. 2, we did obtain beautiful patterns from the vein matrix assemblages by using synchrotron radiation X-ray diffraction technique. These patterns clearly show sharp diffraction peaks of majorite–pyrope garnet, FeNi metal, and troilite, and the resolution of diffraction lines was high enough that almost no overlapping of diffraction lines of these minerals was observed (Fig. 5. 53). All together 16 diffraction reflections were obtained for our majorite–pyrope garnet (Table 5.18). This pattern is similar to that of standard majorite of JCPDS No. 25-0843 (Table 5.18, right column), and our data can be considered as the characteristic X-ray diffraction pattern for the liquidus majorite–pyrope garnet. It is also worth to mention that the diffraction peaks we obtained for majorite–pyrope garnet are sharp enough, and no amorphous phases in any of the probed areas were observed.

TEM Observations

TEM observations conducted on an ion-thinned Suizhou fine-grained vein matrix sample revealed that majorite–pyrope garnet is the most abundant constituent

Table 5.18 Synchrotron radiation X-ray diffraction data of majorite–pyrope garnets in the Suizhou vein matrix (Xie et al. 2011a)

Maj-Prp garnet [#]1 in vein matrix		Maj-Prp garnet [#]2 in vein matrix		Majorite garnet JCPDS No. 25-0843	
d (Å)	I	d (Å)	I	d (Å)	I
2.880	70	2.880	70	2.881	70
2.572	100	2.571	100	2.575	100
2.453	50	2.451	40	2.454	45
2.350	35	2.349	35	2.352	30
2.256	35	2.255	30	2.262	35
2.104	10	2.105	10	2.103	18
2.038	20	2.037	30	2.038	25
1.865	50	1.863	25	1.868	25
1.821	10	1.820	10	1.820	10
1.662	20	1.661	20	1.663	20
1.597	40	1.596	30	1.597	40
1.539	70	1.538	50	1.540	60
1.438	10	1.438	10	1.439	17
1.288	10	1.288	15	1.288	15
1.257	15	1.256	10	1.258	20
1.228	15	1.228	10	1.228	14

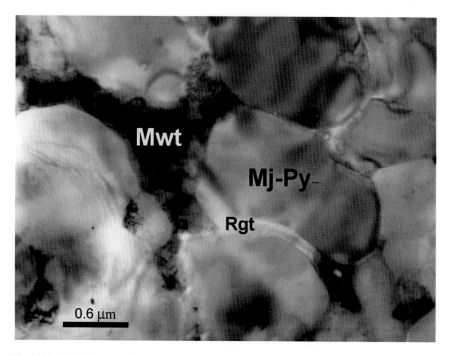

Fig. 5.54 TEM bright field image showing the idiomorphic crystals of majorite–pyrope garnet (*Grt*) and irregular microcrystalline magnesiowüstite (*Mwt*) and ringwoodite (*Rwd*) in interstices between garnet crystals

high-pressure mineral crystallized from silicate melt in shock veins. It occurs in the form of equant idiomorphic crystals (Fig. 5.54). The grain size of such garnet is ranging from 0.3 to 3 μm. The largest garnet crystals can reach to 5 μm in diameter. It is revealed that all garnet crystals show rather smooth surface but slight waving optical features can be observed in some garnet crystals.

Figure 5.55 is the selected area electron diffraction pattern of the Suizhou majorite–pyrope garnet showing the rather regular diffraction spots in cubic symmetry, with space group *Ia3d*, and cell parameter 1.147 nm (Zhang et al. 2006). The appearance of sharp rounded diffraction spots implies that majorite–pyrope garnets in the Suizhou shock veins were crystallized from silicate melt under a rather steady confining high pressure.

Formation Mechanism

In the study of the shock melt veins of the Sixiangkou L6 chondrite, Chen et al. revealed that because the low-Ca pyroxene and olivine in unshocked region of the Sixiangkou meteorite contain little Al, Na, and Cr, the majorite–pyrope solid solution could not have formed by direct transformation from these phases or their high-pressure equivalents, but rather crystallized from a melt that was enriched in Na_2O, CaO, and Cr_2O_3 (Chen et al. 1996a). Chen et al. (1996a) also pointed out

Fig. 5.55 Selected area
electron diffraction pattern of
the Suizhou majorite–pyrope
garnet (After Zhang et al.
2006)

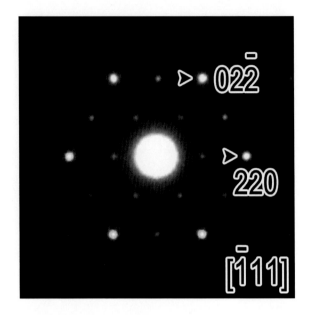

that the shock event on the Sixiangkou L6 chondrite formed a silicate melt of
Sixiangkou bulk composition and that both majorite–pyrope garnets and
magnesiowüstite are crystallized from this melt under very high pressure and
temperature. Since majorite–pyrope garnet in shock veins of the Suizhou meteorite
is also enriched in Na_2O, CaO, and Cr_2O_3, it is evident to maintain that our Suizhou
matrix majorite–pyrope garnet should be a liquidus phase that crystallized from the
shock-induced silicate melt under high pressure and high temperature in the form of
majorite–pyrope solid solution.

P-T Conditions

The garnet- and metal + sulfide-rich matrix assemblage in the Suizhou shock veins
are relatively homogeneous and constitute about 80–90 % of the veins by volume.
Therefore, we assume that this matrix assemblage is originated from the chondritic
material by melting of silicates plus FeNi, FeS, and chromite during a
shock-produced high-pressure and high-temperature event.

On the basis of phase relations in the Mg_2SiO_4–Fe_2SiO_4 and $MgSiO_3$ systems
(Ito et al. 1984; Hogrefe et al. 1994) and high-pressure melting experiments on
peridotite (Zhang and Herzberg 1994) and the Allende carbonaceous chondrite
(Agee et al. 1995), the majorite–pyrope garnet + magnesiowüstite assemblage is
crystallized from 2050 to 2300 °C and 20–24 GPa (Chen et al. 1996a). The lack of
perovskite in the fine-grained assemblage of the veins indicates that the pressure
during crystallization did not exceed about 24 GPa. In the study of the Sixiangkou
shock veins, Chen et al. (1996a) proposed a new model of shock events in meteorites in that the shock-induced high-pressure and high-temperature regime may
retain for up to several seconds. The shock melt veins in Suizhou are extremely

narrow, but the presence of abundant two high-pressure mineral assemblages in veins, the good crystallinity, the lack of amorphous phases, and the idiomorphic forms of the fine-grained majorite–pyrope garnets in the vein matrix suggests a relatively low nucleation rate. Hence, we maintain that the duration of pressure and temperature regime in the Suizhou veins must last, at least, for several seconds.

5.3.2   Magnesiowüstite

At ambient conditions, both FeO (wüstite) and MgO (periclase) adopt the cubic closely packed structure of rock salt. This structure can be envisioned as a stacking of closely packed monolayers along the [111] direction. However, at high pressures, FeO undergoes several phase transitions, whereas MgO does not. The phase believed to occur in the lower mantle is called magnesiowüstite by many investigators, but it actually should be called ferropericlase because all models of the lower mantle assume only 10–20 Fe (Prewitt and Downs 1998). However, magnesiowüstite seems to be the dominate terminology for intermediate compositions and many investigators continue that usage in their publications (Prewitt and Downs 1998). Here, we introduce general characteristics of this mineral found in the Suizhou vein matrix.

Occurrence
Grains of magnesiowüstite in the Suizhou shock veins are very small. It can not be seen by an optical microscope. It is also difficult to be observed under SEM in back-scattered mode. However, the occurrence of magnesiowüstite can clearly be observed by a transmission electron microscope. Under TEM, magnesiowüstite occurs as tiny grains (up to 4 μm long) of irregular form filling the interstitial channels between idiomorphic majorite–pyrope garnet crystals.

Chemical Composition
Compositions of magnesiowüstite in the Suizhou shock veins were measured by EDS technique, and the obtained results are as follows (wt%): MgO—35.45, FeO—62.16, TiO_2—0.20, Cr_2O_3—0.92, SiO_2—0.37, totals—99.10. Its empirical formula is $Wü_{53}$-Per_{47}, where Wü is wüstite and Per is periclase.

Interestingly, the composition of magnesiowüstite in the Suizhou shock veins is quite similar to that of magnesiowüstite in the Sixiangkou shock veins. Both of them contain $\sim$35 wt% of MgO, $\sim$62 wt% of FeO, and $\sim$1 wt% of Cr_2O_3 (Chen et al. 1996a). It was clear that Cr_2O_3 in both Sixiangkou and Suizhou melt was partitioned to magnesiowüstite.

TEM Observations and Electron Diffraction Data
TEM observations revealed that the irregular grains of magnesiowüstite fill the interstices between garnet crystals in the Suizhou vein matrix (Fig. 5.56). It is revealed that the magnesiowüstite grains show polycrystalline nature. Their SAED patterns show sharp diffraction spots of magnesiowüstite (insets of Fig. 4.7a, d), and

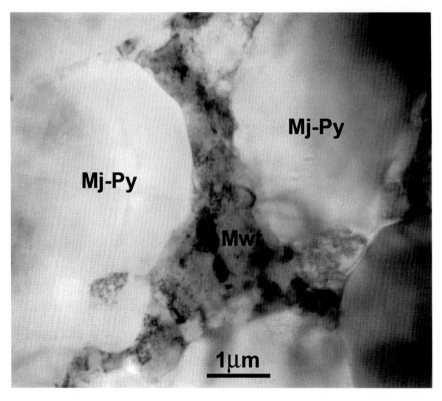

Fig. 5.56 TEM bright field image showing irregular microcrystalline magnesiowüstite (*Mwt*) filling the interstitial channels between idiomorphic majorite–pyrope garnet (*Mj-Py*) crystals

the calculated strong diffraction lines are as follows: 2.120 (100), 1.492 (60), and 2. 439 (40) Å, which can be compared with those of periclase (MgO) in JCPDS No. 4-289 (Table 5.19). The slight increase of *d*-values is due to the presence of FeO in its composition.

Formation Mechanism

Chen et al. (1996a) reported that both majorite–pyrope garnets and magnesiowüstite are crystallized from a silicate melt formed by a shock event on the Sixiangkou L6 chondrite. They revealed that Sixiangkou garnet is enriched in Na_2O, CaO, and Cr_2O_3, and magnesiowüstite is enriched in Cr_2O_3. Similar phenomena were observed in the Suizhou L6 chondrite, e.g., the shock event on the Suizhou L6 chondrite formed a silicate melt of Suizhou bulk composition. Since magnesiowüstite in shock veins of the Suizhou meteorite is also enriched in Cr_2O_3, it is evident that our Suizhou magnesiowüstite should be a liquidus phase that crystallized from the shock-induced silicate melt under high pressure. The location of magnesiowüstite in the interstitial channels between idiomorphic majorite–pyrope garnet crystals indicates that garnet began to crystallize prior to magnesiowüstite.

Table 5.19 Selected area electron diffraction data of fine-grained magnesiowüstite in the Suizhou vein matrix (Xie et al. 2011a)

Fine-grained Mg-wüstite [#]1 in vein matrix		Fine-grained Mg-wüstite [#]2 in vein matrix		Periclase (MgO) JCPDS No. 4-289	
d (Å)	I	d (Å)	I	d (Å)	I
2.439	40	2.437	70	2.431	20
2.120	100	2.116	100	2.106	100
1.492	50	1.499	60	1.489	60
1.273	20	1.275	30	1.270	25
1.216	15	1.218	15	1.216	15
1.065	20	1.065	20	1.055	15
0.968	15	0.963	20	0.966	10
0.946	10	0.945	20	0.942	15
0.887	10	0.886	15	0.860	15
0.8201	10	0.821	10	0.811	3

P-T Conditions

As we mentioned before, majorite–pyrope garnet + magnesiowüstite assemblage is crystallized from silicate melt at 2050–2300 °C and 20–24 GPa (Chen et al. 1996a). The lack of perovskite in the fine-grained assemblage of the Suizhou veins indicates that the pressure during crystallization did not exceed about 24 GPa.

5.3.3   Fine-Grained Ringwoodite Aggregates

As we described earlier in Sect. 5.2.1, ringwoodite, the high-pressure polymorph of olivine, is the main constituent mineral of the coarse-grained assemblage in the Suizhou shock vein. However, we also identified ringwoodite in the fine-grained assemblage in the same Suizhou shock veins. In this section, we shall report the difference in occurrences between these two ringwoodites.

Occurrence

Fine-grained ringwoodite aggregates in the Suizhou vein matrix occur in the form of narrow bands filling the interstitial channels between majorite–pyrope garnet crystals or between garnet and magnesiowüstite grains (Fig. 5.52). They are too small to be observed under the optical microscope or the scanning electron microscope. We identified the fine-grained ringwoodite in the Suizhou vein matrix by using transmission electron microscopic and SAED techniques.

Fig. 5.57 Selected area
electron diffraction pattern of
the fine-grained ringwoodite
in the Suizhou vein matrix

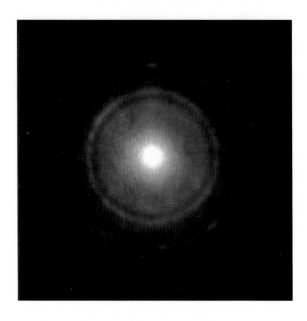

TEM Observations and Selected Area Electron Diffraction Data

Under TEM, three types of microcrystalline ringwoodite bands were revealed,
namely fine granular (Fig. 4.7d), blocky (Fig. 4.7e), and fiber-like (Fig. 4.7f).

The SAED patterns show rather sharp concentric diffraction rings for the
fiber-like type ringwoodite aggregate (Fig. 4.7f, inset image) and concentric
diffraction rings with intermittent diffraction spots for the fine-granular ringwoodite
(Fig. 5.57).

The calculated X-ray diffraction data for the fine-grained ringwoodite are shown
in Table 5.20. The strongest 4 diffraction lines are at 2.446 (100), 2.028 (70), 1.434
(60), and 2.872 (30) Å. They are almost identical to the main reflections of ringwoodite in JCPDS No. 21-1258. This indicates that the crystallinity of the
fine-grained ringwoodite aggregates is still good enough, and no glassy material
was detected in this phase (Xie et al. 2011a).

Formation Mechanism

According to the interstitial occurrence between majorite–pyrope garnet crystals or
between garnet and magnesiowüstite grains, it is believed that the Suizhou
fine-grained ringwoodite in shock veins should also be a liquidus phase that
crystallized from the same Suizhou melt under high pressure after solidification of
majorite–pyrope garnet and magnesiowüstite.

P-T Conditions

The results of high-pressure melting experiments on the Allende carbonaceous
chondrite (Agee et al. 1995) and peridotite (Zhang and Herzberg 1994) indicate that
the majorite–pyrope garnet and magnesiowüstite assemblage are crystallized from
20 to 24 GPa and 2050–2300 °C. The presence of fine-grained ringwoodite rather

Table 5.20 Selected area electron diffraction data of fine-grained ringwoodite in the Suizhou vein matrix (Xie et al. 2011a)

Fine-grained ringwoodite [#]1 in vein matrix		Fine-grained ringwoodite [#]2 in vein matrix		Ringwoodite JCPDS No. 21-1258	
d (Å)	I	d (Å)	I	d (Å)	I
2.872	30	2.879	20	2.872	20
2.446	100	2.445	100	2.447	100
2.028	70	2.027	70	2.028	40
–	–	1.661	5	1.656	<5
1.555	20	1.561	20	1.560	20
1.434	60	1.436	60	1.434	60
1.240	20	1.237	10	1.237	<5
1.170	15	1.176	10	1.172	10
1.021	15	1.015	15	1.014	5
0.901	10	0.901	10	0.907	<5
–	–	0.852	10	0.850	<5
0.827	10	0.828	10	0.828	10

than wadsleyite in the Suizhou vein matrix constrains the pressure to be greater than about 20 GPa. Hence, crystallization of fine-grained ringwoodite would take place at about 20 GPa. This pressure is a bit lower than that for crystallization of the majorite–pyrope garnet and magnesiowüstite assemblage.

5.4   FeNi Metal and Troilite in Shock Melt Veins

Metallic phases in ordinary chondrites are susceptible to the shock metamorphism, both in the solid state and as a result of melting. Shock melting of metal and troilite is observed in ordinary chondrites that have been shocked to shock stage S4 and higher. Melting must have occurred at temperatures >900 °C (Wood 1967; Taylor and Heymann 1971; Smith and Goldstein 1977), and the textural characteristics of the melted metal and sulfide change as the maximum shock temperature increases. In L chondrites of shock stage S4, metal melt droplets are very rare in comparison with troilite droplets and are also much smaller (<2 μm). Fine-grained mixtures of troilite and metal occur at shock stages S4 and above. This texture was thought by Wood (1967) to be the result of eutectic melting. In chondrites of shock stage S6, metal and sulfide melt droplets occur adjacent to most opaque grains and the droplets can reach 50 μm in size. Extensive melting results in the formation of ovoid metal grains (1–200 μm), agglomerated within troilite grains (Taylor and Heymann 1971).

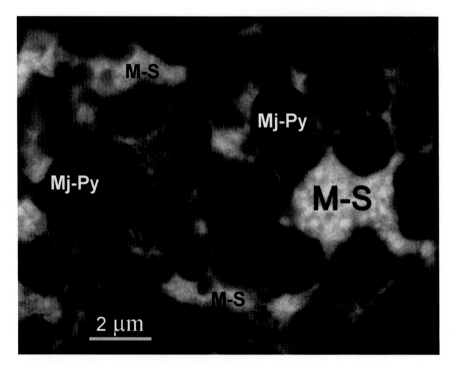

Fig. 5.58 BSE image showing the FeNi metal and troilite eutectic intergrowths (M-S) in the Suizhou shock vein. Note the rounded FeNi metal "islands" (*white*) in the troilite (*gray*) groundmass. Mj-Py = majorite–pyrope garnet in solid solution

Occurrence

In the shock veins of the Suizhou L6 meteorite, FeNi metal and troilite occur mainly in the form of eutectic intergrowths which consist of rounded FeNi metal "islands" in the troilite groundmass (Fig. 5.58). Since the shock melt veins in the Suizhou meteorite are very narrow, the eutectic intergrowths of metal and sulfide in the form of rounded or ellipsoid spherules, or irregular grains, are very small with the diameter less than 2–3 μm. In other cases, molten troilite was injected into shock-induced fractures in silicate or other opaque minerals in the form of opaque melt veins cutting through different minerals in shock melt veins (Fig. 4.2) or forms network of opaque melt veinlets in the Suizhou shock veins.

Chemical Composition

Chemical compositions of FeNi metal and troilite in some metal–sulfide intergrowths are given by Shen and Zhuang (1990) using electron microprobe analysis. For FeNi metal, it gives following results (in wt%): Fe—90.09–91.11, Ni—7.93–7.99, Co—0.44–0.54, Cu—0.00–0.02, Ga—0.06–0.44, Ge—0.29–0.33, S—1.00–1.08 %, and total—99.97–101.45 %. For troilite, the results are (in wt%) as follows: Fe—65.58–78.87, Ni—3.06–14.94, Co—0.00–0.89, Cu—0.00–0.32, Ga—0.00–0.58, Ge—0.06–0.45, S—15.04–30.42, and total—97.82–100.59 %. The average

Table 5.21 Chemical composition of FeNi metal + FeS intergrowths in veins (wt%)

Components	Grain 1	Grain 2	Grain 3	Average
Fe	90.26	90.98	92.82	91.35
Ni	6.14	3.41	3.34	4.30
S	3.60	5.61	3.84	4.35
Total	100.00	100.00	100.00	100.00

All values were determined by EDS

composition for troilite is as follows: Fe—72.21, Ni—7.05, Co—0.30, Cu—0.15, Ga—0.09, Ge—0.24, S—15.50, and total—99.54. From above listed data, we can see that FeNi metal in intergrowths contains small amount of S, and troilite contains some amount of Ni.

In the Suizhou shock veins, some FeNi metal + FeS intergrowth grains are too small to distinguish the two constituent FeNi and FeS phases. Table 5.21 shows the chemical composition of three such FeNi metal + FeS intergrowth grains in a Suizhou shock vein. From this table, we can see that these grains consist of Fe, Ni, and S, reflecting the common composition of these two phases.

As it was pointed out in the Chap. 3 that some small rounded FeNi metal grains of 0.1–5 μm in diameter are also observed in the cracks or intersecting joints of shock-induced planar fractures in olivine and pyroxene (Fig. 3.28). The EPMA results show that this type of metal grains has a much lower Fe content (84.58–85.47 wt%) and higher Ni content (14.53–15.42 wt%) than the large unmelted metal grains in the unmelted chondritic rock. This indicates that the small rounded metal grains are taenite and they formed by precipitation of metal during shock-induced high-temperature melting or even evaporation that might have caused chemical fractionation of Fe and Ni in these metal grains.

Synchrotron Radiation X-ray Diffraction Data

FeNi metal and troilite in some metal–sulfide intergrowths that we observed in the Suizhou shock veins are extremely small. The largest intergrowths are only 2 μm × 3 μm in sizes. They are surrounded all round and underneath by other vein matrix minerals, namely majorite–pyrope garnet, magnesiuwüstite, and ringwoodite (Fig. 5.52). Therefore, the synchrotron radiation X-ray diffraction in situ analysis was used to obtain X-ray diffraction data for FeNi metal and troilite in intergrowths, and the following data were obtained (Fig. 5.53): for FeNi metal: 2.026 (100), 1.453 (20), and 1.171 (10); for troilite: 2.091 (100), 2.99 (60), 2.667 (50), 1.934 (20), 1.782 (20), and 1.332 (10). The above X-ray diffraction data obtained for the FeNi metal and troilite in the Suizhou shock veins are in good consistence with those for the corresponding standard minerals (JCPDS No. 37-0474 for kamacite, and JCPDS No. 37-0477 for troilite).

It should be mentioned that the diffraction peaks we obtained for both FeNi metal and troilite are sharp enough, and no amorphous phases in any of the probed areas were found. This implies that the crystallinity of these two metallic minerals in the Suizhou shock veins is also good enough no matter how fine their grains are.

Fig. 5.59 TEM micrograph
showing the occurrence of
FeNi metal and FeS
intergrowths (M-S).
Mj-Py = majorite–pyrope
solid solution,
Mwt = magnesiowüstite

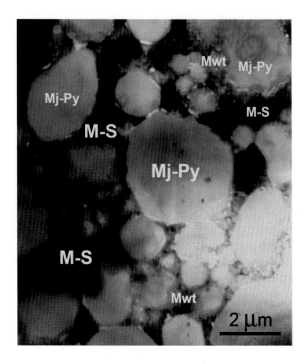

TEM Observations

Since the grains of FeNi metal + FeS intergrowths are not transparent under TEM, it
is hard to explore the microstructure of these grains. The only information that we
could obtained is the dark grain configurations (Fig. 5.59). From this figure, as well
as the SEM BSE image (Fig. 5.58), we can see that the FeNi metal + FeS intergrowth grains are very small, and they really occur in the interstitial channels
between the idiomorphic crystals of majorite–pyrope garnet.

Formation Mechanism and *P-T* Conditions

In the Suizhou shock veins, FeNi metal and FeS occur in the form of fine-grained
intergrowths. This texture was thought to be the result of eutectic melting (Wood
1967). Since the shock melt veins in the Suizhou meteorite are extremely thin, the
size of metal and sulfide melt droplets is also very small, usually only a few μm in
size.

According to the experiments, eutectic melting of FeNi metal and FeS occurs at
temperatures lower than 900 °C at normal pressure (Wood 1967; Smith and
Goldstein 1977). This indicates that the FeNi metal and FeS intergrowths, being the
last crystallization product in the Suizhou shock veins, solidified after pressure
release when the vein temperature decreases lower than 900 °C.

5.5 Summary

(1) Almost all minerals, including silicate, phosphate, and oxide minerals in the Suizhou shock veins, have transformed to their high-pressure polymorphs. The only one mineral that we could not find its high-pressure polymorph is ilmenite.

(2) Ten high-pressure minerals were identified in the shock melt veins of the Suizhou meteorite in two assemblages: One is the coarse-grained assemblage which is consisted of ringwoodite, majorite, akimotoite, vitrified perovskite, lingunite, tuite, xieite, and the CF phase; and another is the fine-grained assemblage which is consisted of matrix minerals majorite–pyrope garnet in solid solution, magnesiowüstite, and microcrystalline ringwoodite.

(3) These two high-pressure mineral assemblages constitute up to 90 % of materials in veins by volume. The other 10 % constituents of veins are fine-grained metal–troilite eutectic intergrowths and blebs which fill the interstices of majorite–pyrope garnet or in the form of metal–sulfide veinlets.

(4) Besides the monomineralic fragments of high-pressure phases in the Suizhou shock veins, some polymineralic fragments of different high-pressure phases are also observed, namely the unique zonal xiete + CF phase + chromite fragments and perovskite + akimotoite + low-Ca pyroxene fragments in adjacent to the vein walls and the three-phase fragments of majorite + vitrified perovskite + ringwoodite, as well as many other two-phase fragments of xieite + one of the high-pressure silicate minerals in the shock veins.

References

Agee CB, Li J, Shannon MC et al (1995) Pressure-temperature phase diagram for Allende meteorite. J Geophys Res 100:17725–17740

Akaogi M (2000) Clues from a shocked meteorite. Science 287:1602–1603

Akaogi M, Hamada Y, Suzuki T et al (1999) High pressure transitions in the system $MgAl_2O_4$–$CaAl_2O_4$: a new hexagonal aluminous phase with implication for the lower mantle. Phys Earth Planet Int 115:67–77

Anderson DL, Bass JD (1986) Transition region of the Earth's upper mantle. Nature 320:321–328

Andrault D, Bolfan-Casanova N (2001) High-pressure phase transformation in the $MgFe_2O_4$ and Fe_2O_3–$MgSiO_3$ system. Phys Chem Mineral 28:211–217

Bell MS, Thomas-Keprta K, Wentworth SJ et al (1999) Pyroxene glass in ALH 84001. Lunar Planet Sci 30:CD-ROM, Abstract #1951

Bertaut EF, Blum P (1956) Determination delta structure de Ti_2CaO_4 par la méthode self-consistante d'approache directe. Acta Crystal 9:121–126

Beswick AE, Carmichael ISE (1978) Constraints on mantle source compositions imposed by phosphorous and the rare-earth elements. Contrib Mineral Petrol 67:317–330

Binns RA (1970) $(Mg,Fe)_2SiO_4$ spinel in a meteorite. Phys Earth Planet Inter 3:156–160

Binns RA, Davis RJ, Read SJB (1969) Ringwoodite, a natural $(Mg,Fe)_2SiO_4$ spinel in the Tenham meteorite. Nature 221:943–944

Bowden KE (2002) Effects of loading path on the shock metamorphism of porous quartz: an experimental study. Ph.D. thesis, University College, London, p 228

Bright NFH, Rowland JF, Wurm JG (1958) The compound CaO · Ti$_2$O$_3$. Can J Chem 36:492–495

Buchwald VF (1984) Phosphate minerals in meteorites and lunar rocks. In: Nriagu JO, Moore PB (eds) Phosphate minerals. Springer, Berlin, pp 199–214

Chen M, Xie XD (1993) The shock effects of olivine in the Yanzhuang chondrite. Acta Mineral Sin 13(2):109–114

Chen M, Xie XD (2015) Shock-produced akimotoite in the Suizhou L6 chondrite. Sci China: Earth Sci 58:876–880

Chen M, Wopenka B, Xie XD et al (1995a) A new high-pressure polymorph of chlorapatite in the shocked chondrite Sixiangkou(L6). Lunar Planet Sci 26:237–238

Chen M, Xie XD, El Goresy A (1995b) Nonequilibrium solidification and micro-structures of metal phases in the shock-induced melt of the Yanzhuang (H6) chondrite. Meteoritics 30:28–32

Chen M, Sharp TG, El Goresy A et al (1996a) The majorite–pyrope + magnesiowüstite assemblage: constrains on the history of shock veins in chondrites. Science 271:1570–1573

Chen M, Wopenka B, El Goresy A (1996b) High-pressure assemblage in shock melt vein in Peace River (L6) chondrite: compositions and pressure-temperature history. Meteoritics 31(Suppl.): A27

Chen M, Xie XD, El Goresy A et al (1998) Cooling rates in the shock veins of chondrites: constraints on the (Mg,Fe)$_2$SiO$_4$ polymorph transformations. Sci China, Series D 41:522–552

Chen M, Shu JF, Xie XD et al (2003a) Natural CaTi$_2$O$_4$-structured FeCr$_2$O$_4$ polymorph in the Suizhou meteorite and its significance in mantle mineralogy. Geochim Cosmochim Acta 67 (20):3937–3942

Chen M, Shu J, Mao HK et al (2003b) Natural occurrence and synthesis of two new postspinel polymorphs of chromite. Proc Natl Acad Sci USA 100(25):14651–14654

Chen M, El Goresy A, Frost D et al (2004a) Melting experiments of a chondritic meteorite between 16 and 25 GPa: implications for Na/K fractionation in a primitive chondritic Earth's mantle. Eur J Miner 16:201–211

Chen M, Xie XD, El Goresy A (2004b) A shock-produced (Mg,Fe)SiO$_3$ glass in the Suizhou meteorite. Meteor Planet Sci 39:1797–1808

Chen M, El Goresy A, Gillet P (2004c) Ringwoodite lamellae in olivine: clue to olivine-ringwoodite phase transition mechanisms in shocked meteorites and in subducting slabs. Proc Natl Acad Sci USA 101:15033–15037

Chen M, Shu JF, Mao HK (2008) Xieite, a new mineral of high-pressure FeCr$_2$O$_4$ polymorph. Chin Sci Bull 53:3341–3345

Coleman LC (1977) Ringwoodite and majorite in the Catherwood meteorite. Can Mineral 15:97–101

Dowty E (1977) Phosphate in Angra Dos Reis: structure and composition of the Ca$_3$(PO$_4$)$_2$ minerals. Earth Planet Sci Lett 35:347–351

Engvik AK, Golla-Schindler U, Berndt J et al (2009) Intragranular replacement of chlorapatite by hydroxy-fluor-apatite. Lithos 112:236–246

Fei Y, Wang Y, Finger LW (1996) Maximum solubility of FeO in (Mg,Fe)SiO$_3$-perovskite as a function of temperature at 26 GPa: implication for FeO content in the lower mantle. J Geophys Res 101:11525–11530

Fei Y, Frost DJ, Mao HK et al (1999) In situ structure determination of the high-pressure phase of Fe$_3$O$_4$. Am Mineral 84:203–206

Ferroir T, Beck P, Van de Moortèle B et al (2008) Akimotoite in the Tenham meteorite: crystal chemistry and high-pressure transformation mechanisms. Earth Planet Sci Lett 275:26–31

Frondel C (1941) Whitlockite, a new calcium phosphate. Am Mineral 26:145

Funamori N, Jeanoz R, Nguyen JH et al (1998) High-pressure transformations in MgAl$_2$O$_4$. J Geophys Res 103:20813–20818

Gasparik P (1992) Melting experiments on the enstatite-pyrope joint at 80–152 kbar. J Geophys Res 97:15181–15188

Gillet P, Chen M, Dubrovinsky L et al (2000) Natural $NaAlSi_3O_8$ –hollandite in the Sixiangkou meteorite. Science 287:1633–1637

Goltrant O, Cordier P, Doukhan JC (1991) Planar deformation features in shocked quartz: a transmission electron microscopy investigation. Earth Planet Sci Lett 106:103–155

Gopal R, Calvo C, Ito J, Sabine WK (1974) Crystal structure of synthetic Mg-whitlockite, $Ca_{18}Mg_2H_2(PO_4)_4$. Can J Chem 52:1152–1155

Graudefroy C, Jouravsky C, Permingeat F (1963) La marokite, $CaMn_2O_4$, une nouvelle espèce minerale. Bull Soc Franç Mineral Crystallogr 86:359–367

Griffin WL, Åmli R, Heier KS (1972) Whitlockite and apatite from lunar rock 14310 and from ÖdegÅrden, Norway. Earth Planet Sci Lett 15:53–68

Hogrefe A, Rubie DC, Sharp TG et al (1994) Metastability of enstatite in deep subducting lithosphere. Nature 372:351–353

Irifine T, Fujino K, Ohtani E (1991) A new high-pressure form of $MgAl_2O_4$. Nature 349:409–411

Irifune T, Ringwood AE, Hibberson WO (1994) Subduction of continental crust and terrigeneous and pelagic sediments: an experimental study. Earth Planet Sci Lett 126:351–368

Ito E, Takahashi E, Matsui Y (1984) The mineralogy and chemistry of the lower mantle: an implication of the ultrahigh-pressure phase relation in the system of MgO-FeO-SiO_2. Earth Planet. Sci. Lett 67:238–248

Jolliff BL, Freeman JJ, Wopenka B (1996) Structural comparison of lunar, terrestrial, and synthetic whitlockite using laser Raman microprobe spectroscopy. Lunar Planet Sci 27:613–614

Kato T, Kumasuwa M (1985) Melting experiment on natural lherzolite at 20 GPa: formation of phase B coexisting with garnet. Geophys Res Lett 13:181–184

Kawai N, Tachimori M, Ito E (1974) A high-pressure hexagonal form of $MgSiO_3$. Proc Jpn Acad 50:378–380

Kerschofer L, Sharp TG, Rubie DC (1996) Intracrystalline transformation of olivine to wadsleyite and ringwoodite under subduction zone conditions. Science 274:79–81

Kerschofer L, Dupas C, Liu M et al (1998) Polymorphic transformation between olivine, wadsleyite and ringwoodite: mechanism of intracrystalline nucleation and the role of elastic strain. Mineral Mag 62:617–638

Kerschofer L, Rubie DC, Sharp TG et al (2000) Kinetics of intracrystalline olivine-ringwoodite transformation. Phys Earth Planet Inter 121:59–76

Kimura M, Suzuki A, Kondo T et al (2000) The first discovery of high-pressure polymorphs, jadeite, hollandite, wadsleyite and majorite, from an H-chondrite, Y-75100. Antarct Meteor 25:41–42

Langenhorst F, Poirier JP (2000) 'Ecologitic' minerals in a shocked basaltic meteorite. Earth Planet Sci Lett 176:259–265

Langenhorst F, Joreau P, Doukhan JC (1995) Thermal and shock metamorphism of the Tenham chondrite: a TEM examination. Geochim Cosmochim Acta 59:1835–1845

Liu LG (1974) Silicate perovskite from phase transformations of pyrope-garnet at high pressure and temperature. Geophys Res Lett 1:277–280

Liu LG (1976) The high-pressure phases of $MgSiO_3$. Earth Planet Sci Lett 31:200–208

Liu LG (1978) High-pressure phase transformations of albite, jadeite and nepheline. Earth Planet Sci Lett 37:438–444

Liu LG, El Goresy A (2007) High-pressure phase transitions of the feldspars and further characterization of lingunite. Int Geol Rev 49:854–860

Malavergne V, Guyot F, Benzerara K et al (2001) Description of new shock-induced phases in the Shergotty, Zagami, Nakhla and Chassigny meteorites. Meteor Planet Sci 36:1297–1305

Mao HK, Takahashi T, Bassett WA et al (1974) Isothermal composition of magnetite to 320 kbar and pressure-induced phase transformation. J Geophys Res 79:1165–1170

Mao HK, Shen G, Hemley RJ (1997) Multivariant dependence of Fe–Mg partitioning in the lower mantle. Science 278:2098–2100

Mason B (1966) Composition of the earth. Nature 21:616–618

Mason B, Nelen J, White JS Jr (1968) Olivine-garnet transformation in a meteorite. Science 160:66–67

McCammon CA (1998) Crystal chemistry of ferric iron in $Fe_{0.05}Mg_{0.95}SiO_3$ perovskite as determined by Mossbauer spectroscopy in the temperature range 80–293 K. Phys Chem Miner 25:292–300

McMillan P (1984a) Structural studies of silicate glasses and melts; application and limitations of Raman spectroscopy. Am Mineral 69:622–644

McMillan P (1984b) A Raman spectroscopic study of glasses in the system $CaO-MgO-SiO_2$. Am Mineral 69:645–659

McMillan P, Akaogi M (1987) Raman spectra of β-Mg_2SiO_4 (modified spinel) and γ-Mg_2SiO_4 (spinel). Am Mineral 72:361–364

Miyagima N, El Goresy A, Dupas-Bruzek C, Seifert F, Rubie BC, Chen M, Xie XD (2007) Ferric iron in Al-bearing akimotoite coexisting with iron-nickel metal in a shock-melt vein. Am Mineral 92:1545–1549

Mosenfelder JL, Marton FC, Ross CR et al (2001) Experimental constraints on the depth of olivine metastability in subducting lithosphere. Phys Earth Planet Inter 127:165–180

Murayama JK, Kato M, Nakai S (1986) A dense polymorph of $Ca_3(PO_4)_2$: a high pressure phase of apatite decomposition. Phys Earth Planet Inter 44:293–303

Nash WP (1984) Phosphate minerals in terrestrial igneous and metamorphic rocks. In: Nriagu JO, Moore PB (eds) Phosphate minerals. Springer, Berlin, pp 215–241

Presnall DC, Gasparik T (1990) Melting of enstatite ($MgSiO_3$) from 10 to 16.5 GPa and the forsterite (Mg_2SiO_4)—majorite ($MgSiO_3$) eutectic at 16.5 GPa: implications for the origin of the mantle. J Geophys Res 90:15771–15777

Prewitt CT (1975) Meteoritic and lunar whitlockites. Lunar Planet Sci 6:647–648

Prewitt CT, Downs RT (1998) High-pressure crystal chemistry. Rev Miner 37:287–317

Price GD, Putnis A, Agrell SO (1979) Electron petrography of shock-produced veins in the Tenham chondrite. Contrib Miner Petrol 71:211–218

Putnis A, Price GD (1975) High-pressure (Mg, Fe) Si_2O_4 phases in the Tenham chondritic meteorite. Nature 280:217–218

Reid AF, Ringwood AE (1969) Newly observed high pressure transformation in Mn_3O_4, $CaAl_2O_4$, and $ZrSiO_4$. Earth Planet Sci Lett 6:205–208

Reid AF, Ringwood AE (1970) The crystal chemistry of dense M_3O_4 polymorph: High pressure Ca_2GeO_4 of K_2NiF_4 structure type. J Solid State Chem 1:557–565

Ringwood AE (1962) Mineralogical constitution of the deep mantle. J Geophys Res 67:4005–4010

Ringwood AE, Major A (1966) Synthesis of Mg_2SiO_4–Fe_2SiO_4 solid solutions. Earth Planet Sci Lett 1:241–245

Ringwood AE, Reid AF, Wadsley AD (1967) High-pressure $KAlSi_3O_8$, an aluminosilicate with sixfold coordination. Acta Cryst 23:1093–1095

Roux P, Louër D, Bonel G (1978) Sur nouvelle forme crystalline du phosphate tricalcique. C.R. Acad Sci Paris Ser C 286:549–551

Rubin AE (1997) Mineralogy of meteorite groups. Meteor Planet Sci 32:231–247

Saxena SK, Dubrovinsky LS, Lazor P et al (1996) Stability of perovskite ($MgSiO_3$) in the Earth's mantle. Science 274:1357–1359

Scarfe CM, Mysen BO, Rai CS (1979) Invariant melting behavior of mantle material: partial melting of two lherzolite nodules. Carnegie Inst Wash Yearb 78:498–501

Serghiou G, Zerr A, Boehler R (1998) (Mg,Fe)SiO_3-perovskite stability under lower mantle conditions. Science 280:2093–2095

Sharp TG, DeCarli PS (2006) Shock effects in meteorites. In: Binzel RP (ed) Meteorites and the early solar system II. The University of Arizona Press, Tucson, pp 653–677

Sharp TG, Chen M, El Goresy A (1997a) Mineralogy and microstructures of shock-induced melt veins in the Tenham (L6) chondrite. Lunar Planet Sci 28:1283–1284

Sharp TG, Lingemann CM, Dupas C et al (1997b) Natural occurrence of $MgSiO_3$-ilmenite and evidence for $MgSiO_3$–perovskite in a shocked L chondrite. Science 277:255–352

Shen SY, Zhuang XL (1990) A study of the opaque minerals and structural characteristics of the Suizhou meteorite. In: A synthetical study of Suizhou meteorite. Publishing House of the China University of Geosciences, Wuhan, 40–52 (in Chinese)

Smith BA, Goldstein JI (1977) The metallic microstructures and thermal histories of severely reheated chondrites. Geochim Cosmochim Acta 41:1061–1072

Smith JV, Mason B (1970) Pyroxene-garnet transformation in Coorara meteorite. Science 168:822–823

Sugiyama S, Tokonami M (1987) Structure and crystal chemistry of a dense polymorph of tricalcium phosphate $Ca_3(PO_4)_2$: a host to accommodate large lithophile elements in the earth's mantle. Phys Chem Miner 15:125–130

Taylor GJ, Heymann D (1971) Postshock thermal histories of reheated chondrites. J Geophys Res 76:1879–1893

Tomioka N, Fujino K (1997) Natural $(Mg,Fe)SiO_3$-ilmenite and -perovskite in the Tenham meteorite. Science 277:1084–1086

Tomioka N, Fujino K (1999) Akimotoite, $(Mg, Fe) SiO_3$, a new silicate mineral of the ilmenite group in the Tenham chondrite. Am Mineral 84:267–271

Tomioka N, Kimura M (2003) The breakdown of diopside to Ca-rich majorite and glass in a shocked H chondrite. Earth Planet Sci Lett 208:271–278

Tomioka N, Mori H, Fujino K (2000) Shock-induced transition of $NaAlSi_3O_8$ feldspar into a hollandite structure in a L6 chondrite. Geophys Res Lett 27:3997–4000

Tutti F, Dubrovinsky LS, Saxena SK (2000) High pressure phase transformation of jadeite and stability of $NaAlSiO_4$ with calcium-ferrite type structure in the lower mantle conditions. Geophys Res Lett 27:2025–2028

Villiers PR, Herbstein FH (1968) Second occurrence of marokite. Am Mineral 53:495–496

Wang DD (1993) An introduction to Chinese meteorites. Science Press, Beijing, 101–106 (in Chinese)

Wang WY, Takahashi E (2000) Subsolidus and melting experiments of K-doped peridotite KLB-1 to 27 GPa: its geophysical and geochemical implications. J Geophys Res 105:2855–2868

Wang Z, Lazor P, Saxena SK et al (2002) High-pressure Raman spectroscopic study of spinel $(ZnCr_2O_4)$. J Solid State Chem 165:165–170

Wood JA (1967) Chondrites: their metallic minerals, thermal histories, and parent planets. Icarus 6:1–49

Xie XD, Chao ECT (1987) Studies on the lattice distortion and substructures of shock lamellae in naturally shocked quartz. Chin J Geochem 6:19–32

Xie XD, Chen M (2009) Shock melting and fractional crystallization of meteorite minerals under dynamic high-pressures and their geochemical significance. Earth Sci Front 16:134–145 (in Chinese with English Abstract)

Xie XD, Chen M, Wang DQ (2001a) Shock-related mineralogical features and P-T history of the Suizhou L6 chondrite. Eur J Miner 13(6):1177–1190

Xie XD, Chen M, Wang DQ et al (2001b) $NaAlSi_3O_8$-hollandite and other high-pressure minerals in the shock melt veins of the Suizhou L6 chondrite. Chin Sci Bull 46(13):1121–1126

Xie XD, Minitti ME, Chen M et al (2002) Natural high-pressure polymorph of merrillite in the shock vein of the Suizhou meteorite. Geochim Cosmochim Acta 66:2439–2444

Xie XD, Minitti ME, Chen M et al (2003) Tuite, γ-$Ca_3(PO_4)_2$, a new phosphate mineral from the Suizhou L6 chondrite. Eur J Mineral 15:1001–1005

Xie XD, Shu JF, Chen M (2005a) Synchrotron radiation X-ray diffraction in situ study of fine-grained minerals in shock veins of Suizhou meteorite. Sci China, Ser D 48:815–821

Xie XD, Chen M, Wang DQ (2005b) Two types of silicate melts in naturally shocked meteorites. Papers and abstracts of the 5th annual meeting of IPACES, Guangzhou, pp 12–14

Xie ZD, Sharp TG (2004) High-pressure phases in shock-induced melt veins of the Umbarger L6 chondrite: Constraints of pressure. Meteor Planet Sci 39:2043–2054

Xie ZD, Sharp TG, DeCarli PS (2006) High-pressure phases in a shock-induced melt vein of the Tenham L6 chondrite: constraints on shock pressure and duration. Geochim Cosmochim Acta 70:504–515

Xie XD, Chen M, Wang CW (2011a) Occurrence and mineral chemistry of chromite and xieite in the Suizhou L6 chondrite. Sci China Earth Sci 54:1–13

Xie XD, Sun ZY, Chen M (2011b) The distinct morphological and petrological features of shock melt veins in the Suizhou L6 chondrite. Meteor Planet Sci 46:459–469

Xie XD, Zhai SM, Chen M, Yang HX (2013) Tuite, γ-$Ca_3(PO_4)_2$, formed by chlorapatite decomposition in a shock vein of the Suizhou L6 chondrite. Meteor Planet Sci 48:1515–1523

Xue X, Zhai S, Kanzaki M (2009) Si-Al distribution in high-pressure $CaAl_4Si_2O_{11}$: A ^{29}Si and ^{27}Al NMR study. Am Mineral 94:1739–1742

Yagi Y, Suzki T, Akaogi M (1994) High pressure transitions in the system $KAlSi_3O_8$–$NaAlSi_3O_8$. Phys Chem Mineral 21:12–17

Yamada H, Matsui Y, Ito E (1984) Crystal-chemical characterization of $KAlSi_3O_8$ with the hollandite structure. Mineral J (Jpn) 12:29–34

Zhang J, Herzberg C (1994) Melting experiments on anhydrous peridotite KLB-1 from 5.0 to 22.5 GPa. J Geophys Res 99:17729–17742

Zhang J, Ko J, HazenR M et al (1993) High-pressure crystal chemistry of $KAlSi_3O_8$ hollandite. Am Mineral 78:493–499

Zhang K, Wang JB, Wang RH (2006) Electron microscopic studies of the mineral phases in the Suizhou meteorite. J Chin Electr Microsc Soc 25(Suppl.):354–355 (in Chinese)

Chapter 6
Shock-Induced Redistribution of Trace Elements

6.1 General Remarks

The statistics of mineral contents in the Suizhou meteorite gives the following results (in volume %): olivine-44, pyroxene-31, plagioclase-14, opaque minerals (FeNi metal, troilite, and chromite)-10, and whitlockite-1 (Zeng 1990). It has also been revealed that the shock melt vein matrix in the Suizhou meteorite is relatively homogeneous and constitutes >80 % of the veins by volume. This vein matrix originated from the chondritic material by melting of silicate minerals, such as olivine, pyroxene, and plagioclase, plus FeNi metal + troilite during a shock-induced high-pressure and high-temperature event (Xie et al. 2001a). This means that the silicate minerals make about 90 % of the vein matrix by volume while the opaque minerals make 10 % of the vein matrix by volume. Hence, we assume that the composition of the Suizhou vein matrix would represent the composition of the Suizhou chondrite itself.

As we know that some larger fragments of silicate minerals olivine, pyroxene, and plagioclase which randomly distributed in the Suizhou vein matrix have transformed into solid state to their high-pressure polymorphs ringwoodite, majorite, and lingunite, respectively, with no change in their chemical compositions (Xie et al. 2001a). However, our detailed LA-ICP-MS studies conducted by first author of this book and Zhang Hong could still reveal some differences in trace element concentrations between high-pressure polymorphs in shock veins and their precursor silicate minerals outside the veins. This implies that some redistribution of trace elements took place during the shock-induced melting and phase transition in the Suizhou melt veins.

X. Xie and M. Chen, *Suizhou Meteorite: Mineralogy and Shock Metamorphism*, Springer Geochemistry/Mineralogy, DOI 10.1007/978-3-662-48479-1_6

6.2   Concentrations of Trace Elements in the Suizhou Minerals

In order to compare the distribution of trace elements in the Suizhou main rock-forming minerals outside shock melt veins with their high-pressure polymorphs inside veins, the REE concentrations on olivine, low Ca-pyroxene, and plagioclase in the Suizhou unmelted chondritic rock, as well as ringwoodite, majorite, and melt vein matrix in veins, were measured by LA-ICP-MS technique. Figures 6.1, 6.2, 6.3, 6.4, and 6.5. Figure 6.6 are BSE images showing the locations of LA-ICP-MS analyses on a polished section of the Suizhou meteorite. Table 6.1 shows the concentrations of trace elements in both silicate minerals and vein matrix melt. For comparison, the trace element concentrations in the Suizhou unmelted chondritic rock analyzed by Wang (1990) and Zhong et al. (1990) using instrumental neutron activation technique are also listed in Table 6.1. From this table, some general trends of redistribution of different types of trace elements can be observed.

6.2.1   Siderophile Elements

Ni and Co in meteorite behave as siderophile elements. The concentrations of Ni and Co in iron meteorites are higher than in any type of chondritic meteorites. It

Fig. 6.1 BSE image showing the location of LA-ICP-MS analysis (*black hole*) of olivine (*Olv*) outside a Suizhou shock vein

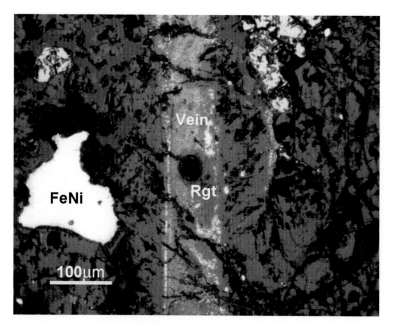

Fig. 6.2 BSE image showing the location of LA-ICP-MS analysis (*black hole*) of ringwoodite (*Rgt*) inside a Suizhou shock vein. FeNi = FeNi metal

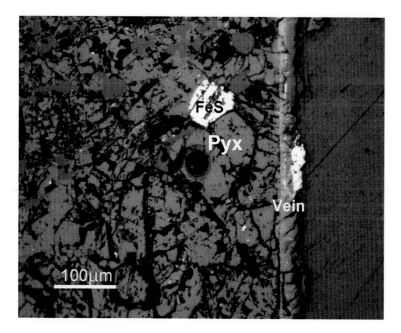

Fig. 6.3 BSE image showing the location of LA-ICP-MS analysis (*black hole*) of pyroxene (*Pyx*) outside a Suizhou shock vein

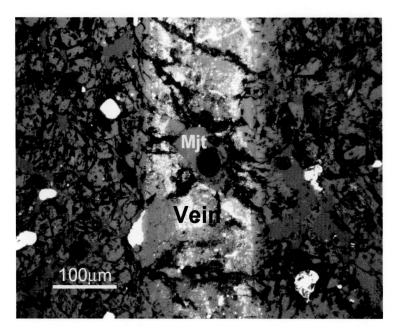

Fig. 6.4 BSE image showing the location of LA-ICP-MS analysis (*black hole*) of majorite (*Mjt*) inside a Suizhou shock vein

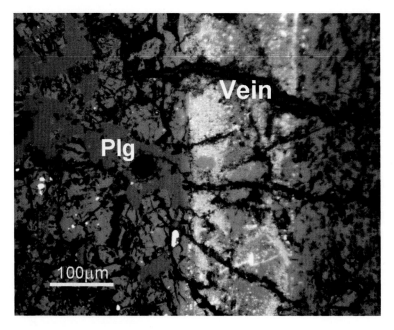

Fig. 6.5 BSE image showing the location of LA-ICP-MS analysis (*black hole*) of plagioclase (*Plg*) outside a Suizhou shock vein

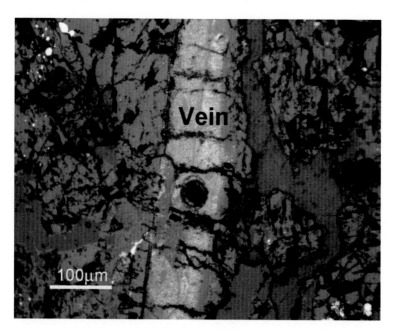

Fig. 6.6 BSE image showing the location of LA-ICP-MS analysis (*black hole*) of vein matrix consisting of majorite-pyrope garnet, magnesiowüstite and FeNi + FeS intergrowths

was found that the variations of Ni and Co concentrations in different types of chondrites are small, but their concentrations increase with the increasing of Fe content in chondrites (Liu et al. 1984). On the other hand, Cr in meteorites behaves as a lithophile and siderophile element, but mainly as a lithophile element. The concentrations of Cr in stone meteorites are much higher than those in iron meteorites. The concentration of V in chondrites (70 ppm in average) is the lowest among the iron group elements, but it is much higher than that in iron meteorites (only a few ppm) (Liu et al. 1984).

The concentrations of siderophile trace elements Ni, Co, Cr, and V in the Suizhou olivine outside veins are relatively low (Ni-5664 ppm, Co-367 ppm, Cr-2302 ppm, and V-175 ppm) (Table 6.1). Our analyses revealed that Ni, Co, and Cr concentrations increased in its high-pressure polymorph ringwoodite (Ni-10630 ppm, Co-762 ppm, and Cr-5814 ppm), i.e., about two times higher than in its precursor olivine (Fig. 6.7a). On the other hand, the content of V in ringwoodite has decreased to 82 ppm, i.e., about one time lower than in olivine.

The contents of siderophile trace elements in the Suizhou pyroxene outside the veins are similar to those in olivine, i.e., Ni-5258 ppm, Co-188 ppm, Cr-1057 ppm, and V-73 ppm. However, when pyroxene transformed to its high-pressure polymorph majorite inside the shock veins, the contents of Ni and Co in majorite have decreased to 2031 ppm and 69 ppm, respectively, i.e., more than two times lower than in its precursor pyroxene (Fig. 6.7b). In contrast to Ni and Co, the contents of

Table 6.1 Contents of trace elements in the Suizhou minerals and melt (ppm)

Element	Olivine (1)	Chondrite Pyroxene (2)	Plagioclase (1)	Ringwoodite (1)	Shock vein Majorite (1)	Vein matrix (4)	Suizhou[a] chondrite (4)
Sc	10.41	1.79	4.96	0.96	0.70	2.17	9.86
V	175.12	72.98	135.20	82.31	960.80	72.98	72.85
Cr	2302.38	1057.0	3546.13	5814.1	69399.4	3577.1	3626
Co	366.73	188.04	945.36	761.65	69.09	1552.35	562
Ni	5664.12	5257.68	14576.74	10629.89	2030.77	27162.87	15600
Cu	25.33	95.94	369.91	219.48	11.739	144.43	–
Zn	57.72	73.07	99.03	134.68	1135.72	107.96	53.2
Ga	10.53	2.93	8.88	15.96	29.57	10.07	5.50
As	3.90	54.41	19.66	5.59	1.57	34.68	1.43
Se	0.00	0.00	0.00	0.00	0.00	14.24	12..57
Sr	29.01	2.21	12.65	6.46	0.99	12.57	12.50
Zr	37.40	4.09	20.89	2.62	1.44	5.37	–
Nb	0.53	0.24	0.98	0.29	0.56	0.60	–
Sb	1.26	0.75	1.93	1.69	0.33	4.20	0.12
Ba	0.00	0.00	0.00	0.00	0.00	0.00	4.62
Y	3.45	0.67	3.45	0.69	0.30	1.90	1.99
La	0.09	0.05	0.46	0.12	0.02	0.28	0.32
Ce	0.21	0.15	0.24	0.21	0.07	0.71	1.37
Pr	0.02	0.00	0.00	0.00	0.00	0.03	–
Nd	9.27	0.60	4.55	2.14	0.33	3.90	1.05
Sm	0.09	0.01	0.30	0.14	0.02	0.34	0.21
Eu	0.59	0.06	0.95	0.37	0.04	0.78	0.08
Gd	0.12	0.01	0.18	0.04	0.01	0.13	0.39
Tb	0.55	0.09	0.79	0.33	0.04	0.80	0.08

(continued)

Table 6.1 (continued)

Element	Olivine (1)	Chondrite Pyroxene (2)	Plagioclase (1)	Ringwoodite (1)	Shock vein Majorite (1)	Vein matrix (4)	Suizhou[a] chondrite (4)
Dy	0.40	0.02	0.23	0.03	0.00	0.22	–
Ho	0.16	0.16	0.15	0.06	0.00	0.14	–
Er	0.06	0.15	0.59	0.22	0.00	0.23	–
Tm	0.06	0.01	0.08	0.01	0.00	0.05	–
Yb	0.51	0.08	0.60	0.16	0.04	0.28	0.25
Lu	0.15	0.02	0.09	0.02	0.00	0.09	0.04
Ta	0.36	0.14	0.45	0.11	0.06	0.22	–
Au	0.05	0.025	0.04	0.003	0.013	0.033	0.15
Hg	0.50	0.20	0.31	0.07	0.11	0.38	–
Tl	0.14	0.03	0.07	0.02	0.02	0.04	–
Pb	0.05	0.02	0.05	0.01	0.02	0.02	–
Bi	0.08	0.04	0.20	0.09	–	0.03	0.00

[a]Averaged from the data of Wang (1990), Zhong et al. (1990). The data were obtained by instrumental neutron activation analysis

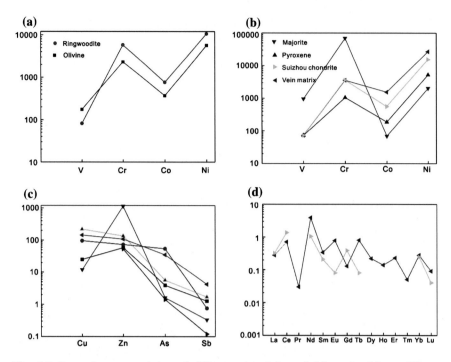

Fig. 6.7 Trace elements variation of different mineral from Suizhou chondrite **a** different characteristics of siderophile elements (V, Cr, Co, and Ni) from ringwoodite and olivine grains; **b** different characteristics of siderophile elements (V, Cr, Co, and Ni) from majorite, pyroxene vein matrix, and Suizhou chondrite grains; **c** different characteristics of chalcophile elements (Cu, Zn, As, and Sb) from each mineral grain; **d** REE distribution patterns from vein matrix and Suizhou chondrite

Cr and V in majorite have greatly increased to 69399 ppm and 961 ppm, respectively, i.e., 65 and 13 times higher than in pyroxene (Table 6.1).

The Suizhou plagioclase contains higher siderophile trace elements (Ni-14577 ppm, Co-945 ppm, and Cr-3546 ppm) than olivine and pyroxene. Since the melting point of plagioclase is much lower than that of olivine and pyroxene, under the shock-induced high pressure and temperature, the molten plagioclase trends to decompose upon shock and the decomposed components transferred into silicate melt, i.e., CaO, Al_2O_3, and Na_2O into crystallized majorite-pyrope garnet, while Ni and Co into FeNi–FeS intergrowths in the vein matrix (Xie et al. 2001a).

The contents of Ni (27163 ppm) and Co (1552 ppm) in the Suizhou vein matrix are higher than in the Suizhou meteorite (Ni-15600 ppm and C0-562 ppm), and much higher than in ringwoodite and majorite embedded in the same vein matrix. On the other hand, the contents of Cr (3577 ppm) and V (73 ppm) in the Suizhou vein matrix are almost identical to those in the Suizhou meteorite (Cr-3626 ppm and V-73 ppm), but lower than in ringwoodite (Cr-5814 and V-82 ppm) and much lower than in majorite (Cr-69399 ppm and V-961 ppm) (Fig. 6.7b, Table 6.1). Since the vein matrix contains about 10 vol. % of FeNi metal + troilite intergrowths, the higher contents of Ni and Co in vein matrix are quite understandable. We assume that the much lower Cr and V contents in the Suizhou vein matrix than in high-pressure silicate minerals are due to the enrichment of these two elements mainly in majorite, where Cr and V may occupy the octahedral M1 site to substitute Al, and partly in ringwoodite, where Mg^{2+} can be replaced by Cr^{3+} or V^{3+}.

6.2.2 Chalcophile Elements

Cu possesses very strong chalcophile property. It is a representative member of the chalcophile element group. The main carrier of Cu in meteorites is troilite. The concentrations of Cu in meteorites show the following trends: metallic minerals > troilite > silicates. It was found that in the conditions of meteorites, the chalcophile property of Cu is close to siderophile property (Liu et al. 1984). It is well known that Pb, Zn, Au, Ag, As, Sb, and Hg belong to the chalcophile element group. Their properties are similar to that of Cu. For instance, the As content in different chondritic minerals shows the same trends as Cu, i.e., metal phase (12 ppm) > sulfide phase (10 ppm) > silicate phase (0.3 ppm). In contrast to Cu and As, the average concentrations of Zn in different meteorites are chondrites (55 ppm) > achondrites (4 ppm) > iron meteorites (<2) (Liu et al. 1984).

We analyzed the concentrations of chalcophile elements, such as Cu, Zn, Pb, As, Sb, Au, Hg, and Bi, in the Suizhou silicate minerals outside and inside the veins, as well as in the melt vein matrix (Fig. 6.7c, Table 6.1). It is found that in comparison with the silicate minerals in the Suizhou meteorite, Cu enriched in vein matrix (144 ppm versus 25.3 ppm in olivine and 96 ppm in pyroxene); element As greatly depleted in majorite (1.57 ppm versus 54.4 ppm in pyroxene) and slightly enriched in vein matrix (34.7 ppm versus 16.0 ppm in ringwoodite and 1.57 ppm in majorite); element Zn enriched in majorite (1136 ppm versus 73.1 ppm in pyroxene) and ringwoodite (135 ppm versus 57.7 ppm in olivine), as well as in vein matrix (108 ppm); and element Sb slightly concentrated in vein matrix (4.2 ppm versus 1.26 ppm in olivine and 0.75 in pyroxene). Our analyses also show that the concentrations of elements Au, Hg, Pb, and Bi in the Suizhou silicate minerals and vein matrix are extremely low (0.1–0.5 ppm). No obvious change of their contents in silicate minerals and vein matrix was observed.

It is clear that the higher Cu concentration in the metal- and troilite-rich vein matrix than that in high-pressure silicate minerals in the Suizhou shock veins is consistent with the above-described trend (FeNi metal > troilite > silicates). Similarly, the depletion of As in silicate mineral majorite and enrichment of As in metal- and troilite-rich vein matrix are controlled by geochemical properties of this element. On the other hand, the enrichment of element Zn in the Suizhou silicate mineral majorite and ringwoodite rather than in the FeNi- and FeS-bearing vein matrix is also controlled by the geochemical properties (substitute Fe^{2+} in silicate minerals during their crystallization from silicate melt) of this element.

6.2.3 Dispersed Elements

According to Liu et al. (1984), the dispersed element Ga exists in all types of meteorites, and its concentration in L-group chondrites is 4.2–8.6 ppm, but its concentration in troilite is extremely low (<0.5 ppm), indicating the weak chalcophile property of this element. The concentrations of Se in ordinary chondrites and iron meteorites are only 4.8–8.2 ppm and 4.6 ppm, respectively, but its concentration in troilite can reach 91–132 ppm. In chondrites, Sr exists mainly in Ca-bearing minerals. Its average content in chondrites is 11 ppm, but much higher content of Sr (75 ppm) is observed in plagioclase. The concentrations of Ga, Se, and Sr in the Suizhou meteorite are 5.50, 12.6, and 12.5 ppm, respectively (Table 6.1). They are in the range of or very close to the average contents in chondritic meteorites.

The concentrations of dispersed elements Ga, Se, Sr, Ba, and Tl in the Suizhou silicate minerals and vein matrix were also analyzed in this study (Table 6.1). It is found that element Ga mainly enriched in majorite (29.6 ppm versus 2.93 ppm in pyroxene) and slightly depleted in ringwoodite (5.85 ppm versus 10.5 ppm in olivine); element Se greatly enriched in the vein matrix (12.2 ppm versus 0 ppm in both olivine and pyroxene). Element Sr not only depleted in ringwoodite (6.46 ppm versus 29.0 ppm in olivine) and majorite (0.99 ppm versus 2.21 ppm a pyroxene), but also depleted in vein matrix (12.6 pm). However, higher concentration of Sr was observed in maskelynite (20.8 ppm).

It is interesting that the Suizhou silicate minerals and the vein matrix do not contain any of Ba (0 ppm in all), and that the concentration of Tl in all the phases is extremely low, but we still can find some depletion of this element in silicate minerals and melt matrix inside the veins. Hence, we assume that except some enrichment of Ga in majorite, all other dispersed elements are partly escaped from silicate minerals olivine and pyroxene during the shock-induced melting and phase transformation in veins. The enrichment of Ga in majorite can be explained by its close relations with Al, the enrichment of Se in vein matrix—by its chalcophile property, and the enrichment of Sr in plagioclase—by its close relations with Ca during the shock metamorphic process.

6.2.4 Rare Elements

The concentration of Zr in the L-group chondrites is in the range of 4.9–19 ppm. In meteorites, Zr behaves as a lithophile element and it is mainly enriched in clinopyroxene (11–110 ppm), and its contents in olivine, orthopyroxene, and plagioclase are relatively low (2–8 ppm) (Liu et al. 1984). The average concentrations of Nb and Ta in meteorites are 0.41 and 0.30 ppm, respectively. They are mainly enriched in silicate phases (Nb-0.50 ppm and Ta-0.38 ppm), and absent in sulfide phases, showing their lithophile properties. The concentrations of other rare elements such as Li, Be, Rb, and Cs in meteorites are extremely low. They all are in the range of ppb to a few ppm (Liu et al. 1984) and exist mainly in silicates.

We only analyzed the concentrations of three rare elements, namely Zr, Nb, and Ta, in the Suizhou silicate minerals and in the melt vein matrix. The results show that the silicate minerals are the main carriers of Zr, although the concentrations of Zr in ringwoodite (2.62 ppm) and majorite (1.44 ppm) are lower than their precursor minerals olivine (37.4 ppm) and pyroxene (4.09 ppm). On the other hand, the concentrations of Nb and Ta in ringwoodite (0.29 and 0.11 ppm) and majorite (0.56 and 0.06 ppm) are rather close to their precursor minerals olivine (0.53 and 0.36 ppm) and pyroxene (0.24 and 0.14 ppm). Similar quantities for Zr, Nb, and Ta are obtained for the shock vein matrix where majorite-pyrope garnet is its predominate component. The above data provide another evidence of that Zr, Nb, and Ta behave as lithophile elements during shock-induced melting in stone meteorites.

6.2.5 Rare Earth Elements

Rare earth elements (REE) in meteorites behave as lithophile elements. The concentrations of REEs in chondrites are in 4 orders of magnitude higher than those in iron meteorites. Liu et al. (1984) reported the concentrations of each REE in L-group chondrites as follows (in ppm): La—0.30–0.61, Ce—0.835–2.12, Pr—0.13–0.16, Nd—0.57–0.79, Sm—0.198–0.28, Eu—0.07–0.09, Gd—0.277–0.40, Tb—0.044–0.068, Dy—0.26–0.42, Ho—0.068–0.089, Er—0.208–0.31, Tm—0.030-0.044, Yb —0.15–0.28, Lu—0.035–0.043, and Y—1.97–2.60 ppm. It is found that no fractionation of REEs appeared not only in chondrites and Ca-achondrites, but also in iron meteorites. It is also revealed that with the increasing of atomic numbers, the REE concentration in meteorites decreases, and the main carriers of REEs in meteorites are Ca-rich minerals plagioclase, pyroxene, whitlockite, and apatite.

The concentrations of most REEs in the Suizhou meteorite are settled in the above-described concentration ranges for L-group chondrites, namely La—0.32 ppm, Ce—1.37 ppm, Sm—0.21, Eu—0.08 ppm, Gd—0.39 ppm, Tb—0.08 ppm, Yb—0.25 ppm, Lu—0.04 ppm, and Y—1.99 ppm (Table 6.1, last column). The only exception is concerned with Nb, where its concentration (1.05 ppm) is slightly higher than the upper limit for this element (0.79 ppm) in L-group chondrites.

It should be pointed out that the concentrations of some REEs in the Suizhou melt vein matrix are similar to those in the Suizhou and other L-group chondrites, for instance, La—0.28 ppm, Sm—0.34 ppm, Dy—0.22 ppm, Er—0.23, Tm—0.05 ppm, Yb—0.28 ppm, Lu—0.09 ppm, and Y—1.90 ppm (Fig. 6.7d, Table 6.1). However, the other REEs show increased or decreased concentrations than those in the Suizhou and other L-group chondrites. Among them, the concentrations of Nd (3.90 ppm), Eu (0.78 ppm), Tb (0.80 ppm), and Ho (0.14 ppm) are higher, and the concentrations of Ce (0.71 ppm), Pr (0.03 ppm), and Gd (0.13 ppm) are lower than in the Suizhou and other L-group chondrites.

The concentrations of REEs in the Suizhou olivine and pyroxene, and their high-pressure polymorphs ringwoodite and majorite, as well as in maskelynite outside melt veins and in the melt vein matrix, are shown in the lower part of Table 6.1. From this table, following REE distribution status can be observed:

1. The REE concentrations in the Suizhou meteorite follow the Oddo–Harkins rule which holds that elements with an even atomic number are more common than elements with an odd atomic number, such as Ce (1.37 ppm) > La (0.32 ppm), Nb (1.05 ppm) > Pr (0.00 ppm), Gd (0.39 ppm) > Eu (0.08 ppm), Yb (0.25 ppm) > Tm (0.00 ppm). The REE concentrations in the Suizhou silicate minerals and in melt vein matrix also follow the Oddo–Harkins rule in general, but with some exceptions. For instance, the concentrations of Gd (Z = 64) and Dy (Z = 66) in all measured Suizhou silicate minerals and in melt vein matrix are lower than Eu (Z = 63) and Tb (Z = 65), respectively.

2. The total REE contents in different mineral phases are olivine (15.83 ppm) > plagioclase (12.49 ppm) > vein matrix (9.88 ppm) > ringwoodite (4.45 ppm) > pyroxene (2.14 ppm) > majorite (0.87 ppm). This indicates that during phase transition of olivine to ringwoodite and pyroxene to majorite, REE contents greatly decreased. We assume that the REEs released from olivine and pyroxene, as well as from plagioclase, might be transferred into Ca-bearing majorite-pyrope garnet in vein matrix, since Ca minerals can enrich REEs during magmatic process (Liu et al. 1984).

3. The ratios of light REEs *vs* heavy REEs in different mineral phases are olivine (10.27 : 5.56 = 1.85) = ringwoodite (2.89 : 1.56 = 1.85) > vein matrix (6.04 : 3.84 = 1.57) > plagioclase (6.51 : 6.07 = 1.07) > majorite (0.48 : 0.39 = 1.23) > pyroxene (0.87 : 1.27 = 0.68). This indicates that all mineral phases are enriched in LREE except pyroxene, where the contents of LREE are higher than HREE.

On the basis of above-described facts, we assume that redistribution of some REEs did take place between different phases during the shock-induced melting and phase transition in the Suizhou veins.

6.3   Summary

1. During the shock-induced melting and phase transition of minerals in the Suizhou melt veins, the siderophile elements such as Ni, Co, Cr, and V partly released from silicate minerals, and then Ni and Co transferred into FeNi + FeS bearing vein matrix, while Cr and V transferred into coarse-grained majorite in melt veins to substitute Al^{3+}.

2. The chalcophile elements such as Cu, As, and Sb are enriched in FeNi + FeS bearing vein matrix, while Zn slightly enriched in vein matrix but greatly enriched in coarse-grained majorite to substitute Fe^{2+}.
3. The dispersed element Ga mainly enriched in majorite for its close relations with Al, and Se greatly enriched in FeNi + FeS bearing vein matrix for its chalcophile property, while Sr is not only depleted in ringwoodite and majorite but also depleted in vein matrix.
4. Silicate minerals are main carriers of rare elements, Zr, Nb, and Ta, and their concentrations are relatively stable during shock melting and phase transition in the Suizhou melt veins.
5. The REE concentrations in the Suizhou L6 chondrite are very low, and the total REE content is only a few ppm. During shock-induced melting and phase transition in the Suizhou melt veins, REEs partly transferred from REE-rich silicate minerals olivine and plagioclase into Ca-bearing majorite-pyrope garnet in vein matrix.

References

Liu YJ, Cao LM, Li ZL et al (1984) Geochemistry of elements. Science Press, Beijing, p 548

Wang DD (1990) Characteristics of the trace element abundance of the Suizhou meteorite. In: A synthetical study of Suizhou meteorite. Publishing House of the China University of Geosciences, Wuhan, pp 77–79 (in Chinese)

Xie XD, Chen M, Wang DQ (2001a) Shock-related mineralogical features and P-T history of the Suizhou L6 chondrite. Eur J Miner 13(6):1177–1190

Xie XD, Chen M, Dai CD et al (2001b) A comparative study of naturally and experimentally shocked chondrites. Earth Planet Sci Lett 187:345–356

Zeng GC (1990) Common transparent minerals and chemico-petrological type of the Suizhou meteorite. In: A synthetical study of Suizhou meteorite. Publishing House of the China University of Geosciences, Wuhan, pp 32–40 (in Chinese)

Zhong HH, Jiang LJ, Yang XH et al (1990) A preliminary study of the micro-elements in the Suizhou meteorite. In: A synthetical studiy of Suizhou meteorite. Publishing House of the China University of Geosciences, Wuhan, pp 74–77 (in Chinese)

Chapter 7
Evaluation of Shock Stage
for Suizhou Meteorite

7.1 General Remarks

Sharp and DeCarli (2006) reported that shock waves have played an important role in the history of virtually all meteorites. It is well known that throughout the history of the solar system, asteroids and other small bodies, such as meteorite parent bodies, have repeatedly collided with one another and with the planets. Since collisions produce shock waves in the colliding bodies, an understanding of shock wave effects is important to unravel the early history of the solar system as it is revealed in meteorites.

Shock-induced deformation features are observed in many chondritic meteorites (Fredriksson et al. 1963; Carter et al. 1968; Binns et al. 1969; Ashworth and Barber 1975; Dodd and Jarosewich 1979; Stöffler et al. 1988, 1991; Chen et al. 1995a, 1996a, b, 2000; Xie et al. 1991, 2001a, b, c, 2011a, b). Detailed studies on these phenomena were carried out in an attempt to quantify these features in terms of progressive stages of shock metamorphism and estimate the peak shock pressures and temperatures (Stöffler et al. 1991; Schmitt and Stöffler 1995). Stöffler et al. (1991) defined a scheme of six stages of shock (S1-S6) on the basis of shock effects in olivine and plagioclase and proposed a shock pressure calibration for these stages on the basis of a critical evaluation of data from shock recovery experiments (Stöffler et al. 1991). The characteristic effects of each shock stage are given as follows: S1 (unshocked)—sharp optical extinction of olivine; S2 (very weakly shocked)—undulatory extinction of olivine; S3 (weakly shocked)—planar fracture in olivine; S4 (moderately shocked)—mosaicism in olivine; S5 (strongly shocked)—isotropization of plagioclase (maskelynite) and planar deformation features in olivine; and S6 (very strongly shocked)—recrystallization of olivine, sometimes combined with phase transformation of olivine (ringwoodite and/or phases produced by dissociation reactions). S6 effects are always restricted to regions adjacent to melted portions of a sample which is otherwise only strongly shocked. These authors also reported that shock melt veins are not critical for the definition of the

© Springer-Verlag Berlin Heidelberg and Guangdong Science
& Technology Press Co., Ltd. 2016
X. Xie and M. Chen, *Suizhou Meteorite: Mineralogy and Shock Metamorphism*,
Springer Geochemistry/Mineralogy, DOI 10.1007/978-3-662-48479-1_7

shock stage even though their presence indicates that the meteorite is shocked to stage 3 or above. Their scheme reveals a progressive increase in the abundance of opaque shock veins (S3), ranging from melt pockets, interconnecting melt veins, and opaque shock veins (S4) to the pervasive formation of melt pockets, melt veins, and opaque shock veins (S5 and S6). Diaplectic plagioclase glass ("maskelynite") appears in stage S5, whereas the high-pressure polymorph of olivine, ringwoodite, is seen in restricted local domains at stage S6.

The Suizhou meteorite is a shocked and melt-vein-bearing L6 chondrite. Shock effects are developed in the Suizhou olivine, pyroxene, and other minerals, as well as in the formation of shock melt veins. According to our previous investigations, the shock features of the Suizhou meteorite match shock stages from S3 to S6 of the scheme defined by Stöffler et al. (1991). However, our recent studies revealed that the proper shock stage for this meteorite would be S5. In this chapter, we summarize the shock-related mineralogical features of the Suizhou meteorite and try to properly evaluate the shock stage for this meteorite.

7.2 General Shock Features of the Suizhou Meteorite

On the basis of shock effects observed in olivine and pyroxene, the Suizhou meteorite was evaluated by previous investigators as a weakly shock-metamorphosed (S2-S3) meteorite (Wang and Li 1990; Wang 1993). The determination of the cosmogenic nuclides ^{3}He, ^{21}Ne, ^{35}Ar, ^{83}Kr, and ^{126}Xe, as well as the ^{81}Kr-Kr dating, indicates that the Suizhou meteorite has an average cosmic-ray exposure age of (29.8 ± 3.5) Ma (Wang 1990). The results obtained by different methods are in good agreement within the limits of experimental errors. This implies that the meteorite has only a single-stage cosmic-ray exposure history. Hence, Wang (1990) assumed that the meteorite parent body broke up as a result of a collision to form the Suizhou meteorite at about 30 Ma ago and this meteorite experienced a shock stage S2-S3.

As we mentioned in Chap. 1, it was also revealed that there is no loss of the cosmogenic nuclide ^{3}He since the exposure of the Suizhou meteorite to cosmic rays soon after its separation from the parent body (Wang 1993). This indicates that the Suizhou meteorite was weakly shocked (S3) and the shock pressure should be below 10 GPa.

Our recent studies of the Suizhou meteorite indicate that abundant irregular and planar fractures are developed in olivine and pyroxene, and most of the plagioclase grains were melted and transformed to maskelynite. Shock melt veins are developed in the Suizhou meteorite, and abundant high-pressure mineral phases were identified in these veins. The boundaries separating the black vein material from the light-colored unmelted chondritic rock are sharp and straight. "Shock blackening" effect and shock-induced melt pockets were not encountered in the Suizhou meteorite.

7.3   Shock Effects in the Suizhou Minerals

Stöffler's scheme of six shock stages (S1-S6) was defined on the basis of shock effects observed in olivine and plagioclase (Stöffler et al. 1991). This is feasible when you focused on the internal structures of these two minerals under the optical microscope, but sometimes one may get incorrect result in the evaluation of shock stage for a meteorite, if the study is not a comprehensive and deep going one. For example, a shock stage of S2-S3 for the Suizhou meteorite was originally evaluated by previous investigators (Wang and Li 1990; Wang 1993) and that of S3-S4 by ourselves (Xie et al. 2001a), but our recent detailed studies on the shock effects of different minerals in the Suizhou meteorite amended the shock stage from S2-S3 and S3-S4 to S5.

7.3.1   Olivine and Pyroxene

Olivine is the most abundant constituent mineral (~ 60 % by volume) in the unmelted domain of the Suizhou meteorite. This mineral displays wavy extinction, undulatory extinction, irregular fractures and from one to four sets of parallel planar fractures with a spacing of tens of micrometers (Figs. 3.1 and 3.2). Some olivine grains show mosaic structure with subgrains of 150–200 μm in sizes (Fig. 3.3). Furthermore, in some parts of the thin sections, the olivine grains are granulated with a grain size less of than 10 μm, but neither planar deformation features (PDF), nor solid state recrystallization were observed in olivine grains. All these effects indicate that the Suizhou olivine was shocked to stage S4, or even to S5 according to Stöffler's classification.

Low-calcium pyroxene in the Suizhou meteorite displays wavy extinction, irregular fractures, and mechanical polysynthetic twinning, with a grain size of 0.05–0.8 mm. Shock-induced mosaic structures and granulation (down to grain sizes of <0.02 mm) are also observed in some pyroxene grains (Fig. 3.21), but no PDF or transformation into molten state was found in the Suizhou pyroxene. From the above description, it is clear that similar to olivine, the low-Ca pyroxene in the Suizhou chondrite is also shocked to stage S4, or even to S5 according to Stöffler's classification.

7.3.2   Plagioclase

Plagioclase is a common rock-forming mineral in L-group chondrites. However, we found that most plagioclase grains in the Suizhou meteorite were melted and transformed to maskelynite. Only a few small grains of untransformed plagioclase are observed in thin sections. The maskelynite, representing a quenched high-pressure melt, displays a smooth surface without any traces of cleavages and scarce fractures (Fig. 3.21).

One of the specific features of the Suizhou maskelynite is the flowing character. Long veinlets and offshoots of maskelynite are commonly observed in the Suizhou chondritic rock (Fig. 3.24). They were formed through injection of plagioclase melt into fractures or cracks in the neighboring olivine and pyroxene. Another specific feature of the Suizhou maskelynite is the formation of plagioclase melt pockets containing tiny chromite and silicate fragments (Fig. 3.46).

According to Stöffler's classification, the appearance of isotropization of plagioclase (maskelynite) in a meteorite corresponds to the shock stage S5. Hence, we assume that our Suizhou meteorite is strongly shocked to stage S5.

7.3.3 FeNi Metal and Troilite

Metal and troilite in the Suizhou meteorite occur as single grains of irregular shape, and display no obvious intragranular textures. Raman spectra taken on troilite grains show rather sharp peaks, indicating that they were weakly shock deformed (<S3). However, some small rounded FeNi metal grains of 0.5–5 μm in diameter were observed in the cracks or intersecting joints of shock-induced planar fractures in pyroxene and olivine (Figs. 3.9 and 3.28). These small rounded metal grains contain much higher Ni content (14.53–15.42 wt%) and much lower Fe content (84.58–85.47 wt%) than the large FeNi metal grains in the Suizhou chondritic rock (4.40–6.81 wt% of Ni and 91.15–94.80 wt% of Fe) (Wang and Li 1990). This indicates that these small rounded FeNi metal grains might have experienced strong shock of stage S5, that causes melting or evaporation of metal phase and subsequent chemical fractionation. Therefore, we consider that this type of metal grains is of secondary origin. They were deposited in cracks and fracture joints from nearby shock-induced Ni-rich metal melt or, more likely, from the Ni-rich metal vapor phase.

7.3.4 High-Pressure Minerals in Shock Melt Veins

As it was described in Chap. 5, the thin shock melt veins in the Suizhou meteorite are full of high-pressure mineral phases. These phases can be divided into two assemblages, namely the fine-grained matrix assemblage consisted of majorite-pyrope solid solution, magnesiowüstite, and microcrystalline ringwoodite; and the coarse-grained assemblage consisted of ringwoodite, majorite, akimotoite, lingunite, tuite, xieite, etc. Stöffler et al. (1991) reported that shock melt veins are not critical for the definition of the shock stage even though their presence indicates that the meteorite is shocked to stage 3 or above. They also emphasized that the pervasive formation of melt veins and opaque shock veins would correspond to shock stages S5 and S6, whereas ringwoodite, the high-pressure polymorph of olivine, is seen in restricted local domains at stage S6. Hence, we assume that shock melt veins in the Suizhou meteorite would surely be shocked to stage S6.

7.4   Summary

On the basis of above-described shock features of olivine, pyroxene, plagioclase, and FeNi metal, it is proper to evaluate the shock stage for the Suizhou unmelted chondritic rock to be S5 and that for the Suizhou shock veins to be S6. Since the Suizhou meteorite contains only a few thin melt veins that makes only <2 % of the meteorite by volume, the unmelted chondritic rock makes up >98 % of the meteorite by volume. It is believed that the whole rock of the Suizhou meteorite could be strongly shocked up to stages S5 with some localized regions (veins) in stage 6.

References

AshworthJ R, Barber DJ (1975) Electron petrology of shock-deformed olivine in stone meteorites. Earth Planet Sci Lett 27:43–54

Binns RA, Davis RJ, Read SJB (1969) Ringwoodite, a natural (Mg, Fe)$_2$SiO$_4$ spinel in the Tenham meteorite. Nature 221:943–944

Carter NL, Raleigh CB, DeCarli PS (1968) Deformation of olivine in stone meteorites. Geophys Res 73:5439–5461

Chen M, El Goresy A (2000) The nature of maskelynite in shocked meteorites: not diaplectic glass but a glass quenched from shock-induced dense melt at high-pressures. Earth Planet Sci Lett 179:489–502

Chen M, Wopenka B, Xie XD et al (1995) A new high-pressure polymorph of chlorapatite in the shocked chondrite Sixiangkou(L6). Lunar Planet Sci 26:237–238

Chen M, Sharp TG, El Goresy A et al (1996a) The majorite—pyrope+magnesiowustite assemblage: Constrains on the history of shock veins in chondrites. Science 271:1570–1573

Chen M, Wopenka B, El Goresy A (1996b) High-pressure assemblage in shock melt vein in Peace River (L6) chondrite: compositions and pressure-temperature history. Meteoritics, 31(Suppl.): A27

Dodd RT, Jarosewich E (1979) Incipient melting and shock classification of L-group chondrites. Earth Planet Sci Lett 44:335–340

Fridriksson K, DeCarli PS, Aaramäe A (1963) Shock-induced veins in chondrites. Space Res 3:974–983

Schmitt RT, Stöffler D (1995) Experimental data in support of the 1991 shock classification of chondrites (Abstract). Meteoritics 30:574–575

Sharp TG, DeCarli PS (2006) Shock effects in meteorites. In: Binzel RP (ed) Meteorites and the early solar system II. The University of Arizona Press, Tucson, pp 653–677

Stöffler D, Bischoff A, Bushward V et al (1988) Shock effects in meteorites. In: Kerridge JF, Matthews MS (eds) Meteorites and the earth solar system. University of Arizona Press, Tucson, pp 165–202

Stöffler D, Keil K, Scott ED (1991) Shock metamorphism of ordinary chondrites. Geochim Cosmochim Acta 55:3854–3867

Wang DD (1993) An introduction to Chinese meteorites. Science Press, Beijing, pp 101–106

Wang DD, Li ZH (1990) A study of rare gas dating of the Suizhou meteorite. In: A synthetical study of Suizhou meteorite. Publishing House of the China University of Geosciences, Wuhan, pp 125–128 (in Chinese)

Xie XD, Li ZH, Wang DD et al (1991) The new meteorite fall of Yanzhuang—a severely shocked H6 chondrite with black molten materials (Abstract). Meteoritica 26:411

Xie XD, Chen M, Wang DQ (2001a) Shock-related mineralogical features and P-T history of the Suizhou L6 chondrite. Eur J Miner 13(6):1177–1190

Xie XD, Chen M, Wang DQ et al (2001b) NaAlSi$_3$O$_8$-hollandite and other high-pressure minerals in the shock melt veins of the Suizhou L6 chondrite. Chin Sci Bull 46(13):1121–1126

Xie XD, Chen M, Dai CD et al (2001c) A comparative study of naturally and experimentally shocked chondrites. Earth Planet Sci Lett 187:345–356

Xie XD, Sun ZY, Chen M (2011a) The distinct morphological and petrological features of shock melt veins in the Suizhou L6 chondrite. Meteor Planet Sci 46:459–469

Xie XD, Chen M, Wang CW (2011b) Occurrence and mineral chemistry of chromite and xieite in the Suizhou L6 chondrite. Sci China Earth Sci 54:1–13

Chapter 8
P-T History of the Suizhou Meteorite

8.1 General Remarks

In Chap. 7, we introduced the petrographic classification of six progressive stages (S1–S6) of shock metamorphism of ordinary chondrites proposed by Stöffler et al. (1991). At the same time, a shock pressure calibration for the six stages based on the critical evaluation of data from shock recovery experiments was also proposed by these authors, which defines the S1/S2, S2/S3, S3/S4, S4/S5, and S5/S6 transitions at <5, 5–10, 15–20, 30–35, and 45–55 GPa, respectively (Stöffler et al. 1991). The fact that olivine and pyroxene do not transform to their high-pressure polymorphs in shock recovery experiments has been used by Stöffler et al. (1991) as evidence that extreme pressure is needed to transform them in meteorites, so the shock pressure range for the reconstructive olivine–ringwoodite transition is thought to be between 45 and 90 GPa and the post-shock temperature between 600 and 1750 °C, and the whole-rock melting and formation of impact melt rocks or melt breccias occurs at about 75–90 GPa (Stöffler et al. 1991).

Shock recovery experiments have duplicated many, but not all, of the metamorphic features observed in meteorites. However, the range of shock condition, such as loading path, peak pressure, peak pressure duration, and unloading path, accessible in laboratory shock recovery experiments is sharply limited in comparison with natural events. Therefore, we contacted a comparative study of experimentally and naturally shocked chondrites to explore the feasibility of applying results of shock recovery experiments to calibrate the shock histories of meteorites (Xie et al. 2001c).

© Springer-Verlag Berlin Heidelberg and Guangdong Science
& Technology Press Co., Ltd. 2016
X. Xie and M. Chen, *Suizhou Meteorite: Mineralogy and Shock Metamorphism*,
Springer Geochemistry/Mineralogy, DOI 10.1007/978-3-662-48479-1_8

8.2   Comparison of Experimentally and Naturally Shocked Meteorites

We have systematically studied the recovered samples of Jilin (H5) chondrite shock-loaded at peak pressures between 12 and 133 GPa. The experimentally induced deformation features were then compared with those observed in naturally shocked H chondrite (Yanzhuang) and L chondrite (Sixiangkou).

8.2.1   Sample Preparation and Shock Experiments

Three meteorites, the Jilin H5, Yanzhuang H6, and Sixiangkou L6 chondrites, were chosen for this investigation. The Jilin chondrite was used to conduct shock-loading experiments and to investigate the experimentally produced shock features, because Jilin is a fresh meteorite fall and has experienced only weak shock metamorphism in space (Xie and Huang 1991). Based on the shock classification of chondrites by Stöffler et al. (1991), the shock pressure it experienced was estimated as <15 GPa, and the average shock level in the Jilin is S2-S3. The Yanzhuang and Sixiangkou chondrites are used to study natural shock effects because the Yanzhuang and Sixiangkou were heavily shock-metamorphosed in extraterrestrial impact events (Xie et al. 1991; Begemann et al. 1992; Chen et al. 1996). The average shock levels in Yanzhuang and Sixiangkou are S6 using the Stöffler's classification.

Shock-loading experiments were performed on the dynamic high-pressure device equipped with a two-stage light-gas gun (ϕ23 mm inner caliber) of the China Southwest Institute of Fluid Physics by Dai and his coworkers (Dai et al. 1991). Experimental specimens were made into ϕ10 mm × 2 mm disks. Each specimen was mounted into the recovery capsule in advance, and then a hypervelocity flying plate impacted the sample capsule, so a plane shock wave produced and shock-loaded the meteorite sample. Various impedance flying plates and the two-impact technique (McQueen et al. 1970; Jing 1986) were used to boost the shock pressure. Lexan, Ly-12 Al, and copper were used as flying plates. Copper and tungsten plates were used as high-impedance materials to reflect the shock wave. Impact velocity of the flying plates was measured by magnetoflyer technique (Kondo et al. 1977). Shock pressure was calculated by the impedance matching method (McQueen et al. 1970). The shock-loading conditions of Jilin samples are shown in Table 8.1. We note that the present investigations are the first with peak pressures much higher than those conducted on the Kurnouv H6 chondrite (Sears et al. 1984; Schmitt and Stöffler 1995). From Jilin samples recovered from the shock experiments, and from the Yanzhuang and Sixiangkou meteorites polished thin sections were made and investigated by a MPV-SP optical microscope, a Hitachi S-3500 N scanning electron microscope (SEM) in BSE mode which is equipped with a Link ISIS 300 X-ray energy dispersive spectrometer (EDS), and a CAMACA SX-51 electron microprobe (EPMA). An XY Dilor multichannel

Table 8.1 Shock-loading conditions of Jilin meteorite sample[a]

No. of shots	Impact velocity (km/s)	Shock pressure (GPa)		Temperature (K)		Flyer material	High-impedance material
		Initial	Peak	Shock	Residual		
1	2.508	8.0	12	336	297	Lexan	Cu
2	2.258	18.2	27	472	386	Ly-12 Al	Cu
3	2.621	39.0	39	624	502	Cu	/
4	2.388	35.0	53	720	550	Cu	W
5	2.691	40.5	78	1120	832	Cu	W
6	2.828	43.4	83	1249	925	Cu	W
7	3.066	48.2	93	1460	1064	Cu	W
8	3.927	68.5	133	2426	1725	Cu	W

[a]Dai et al. (1991)

micro-Raman spectrometer (LRM) in the Ecole Normale Superieure de Lyon, France, was used to identify the mineral phases in the Jilin sample shocked at 83 GPa, and the Renishaw RM-1000 Raman microscopes in the Beijing Institute of Non-ferrous Metals and in the China University of Geosciences (Wuhan) were used for identification of mineral phases in Jilin samples shock-loaded at 83, 93, and 133 GPa. The transmission electron microscopy (TEM) was also used for identification of mineral phases in Yanzhuang and Sixiangkou shock melt veins (Xie et al. 2001c).

8.2.2 Results of Shock Recovery Experiments on Jilin Chondrite

The results of optical, SEM, EDS, and LRM studies on the experimentally shock-recovered samples are as follows (Xie et al. 2001c).

At a peak pressure of 12 GPa, fragmentation and deformation of rock and minerals were produced. Silicates, including olivine, pyroxene, and plagioclase, display strong undulatory extinction and planar fractures.

At 27 GPa, 3–4 sets of planar fractures in olivine and weak mosaicism in olivine and pyroxene were developed. Less than ten percent of plagioclase grains transformed into glass. Evidently, the deformations of silicate minerals in these two shock-loaded Jilin samples were enhanced in comparison with those in unshocked samples.

At 39 GPa, strong mosaicism with subgrains of 12–20 μm in size was produced in olivine and pyroxene. About thirty percent of plagioclase transformed into glass.

At a peak pressure of 53 GPa, the whole rock was severely fragmented or brecciated. Mosaicism in olivine and pyroxene is omnipresent along with plagioclase glass. About sixty percent of plagioclase was transformed into glass. Incipient shock veins and pockets were produced, in which plagioclase, metal, and troilite were melted, whereas olivine and pyroxene were crushed into fine grains of micrometer sizes and partially melted. In the veins and pockets, the molten metal and troilite are usually solidified as droplets or blebs.

At peak pressures of 78–83 GPa, seventy percent of the whole rock was broken into grain sizes <30 μm. All plagioclase was transformed into melt glass, and the birefringence of some olivine and pyroxene becomes very low. On the back-scattered electron (BSE) image of SEM, we observed that plagioclase melt glass penetrates or fills into the cracks or cleavages of olivine and pyroxene (Fig. 8.1a). Shock-produced silicate melt veins and pockets were first formed at this pressure regime. There is textural evidence that all silicates including olivine and pyroxene in the veins and pockets were melted (Fig. 8.1b). Chemical analyses

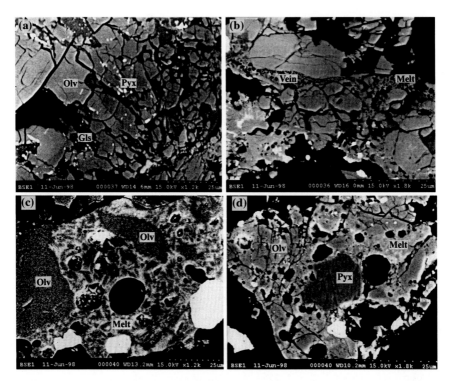

Fig. 8.1 Back-scattered electron images of experimentally shocked Jilin chondrite. **a** Plagioclase melt glass (*Gls*), which formed at a peak pressure of 78 GPa, penetrates into the cracks and cleavages of olivine (*Olv*) and pyroxene (*Pyx*). **b** Shock-produced silicate melt veins (*Vein*) and melt pockets (*Melt*) shocked at a peak pressure of 78 GPa. Vesicles (*black holes*) are observed in the melt regions. **c** A silicate melt pocket (Melt) produced at a peak pressure of 93 GPa is consisting of dendrites of olivine and pyroxene (thin laths with *dark gray color*), and metal and troilite (with *bright gray color*) in interstice. There are some partially melted olivine (*Olv*) fragments in this melt pocket. *Black*-rounded holes in the melt pocket are vesicles. **d** A silicate melt pocket (*Melt*) produced at a peak pressure of 133 GPa contains fragments of olivine (*Olv*) and pyroxene (*Pyx*). *Black holes* in the melt region are vesicles (Xie et al. 2001a)

Table 8.2 Average composition of Jilin meteorite and olivine and pyroxene[a] (wt%)

	Bulk 1	Bulk 2	Olivine			Orthopyroxene		
P (GPa)		83	12	53	83	12	53	83
No.		5	10	10	8	10	7	8
SiO₂	50.28	50.61	38.31	38.87	38.38	56.55	55.57	55.14
TiO₂	0.28	0.16	0.04	0.06	0.03	0.09	0.28	0.09
Al₂O₃	2.81	2.79	n.d	0.03	n.d.	0.11	0.22	0.16
Cr₂O₂	0.54	0.70	0.03	0.06	0.04	0.11	0.12	0.10
FeO	10.06	11.35	17.73	17.65	17.70	11.15	11.14	11.35
MnO	0.39	0.41	0.48	0.47	0.55	0.48	0.57	0.53
MgO	31.35	30.67	42.86	42.20	42.65	31.34	30.73	31.22
CaO	2.55	1.72	0.04	0.05	0.11	0.52	0.54	0.48
K₂O	0.15	0.14	n.d.	n.d.	0.03	n.d.	0.04	0.05
Na₂O	1.15	0.38	n.d.	0.02	0.02	0.03	0.10	0.03
Totals	100.0	98.93	99.99	99.41	99.51	100.38	99.31	99.15
Fa or Fs (mol)			18.2	19.0	18.9	16.5	16.8	16.9

[a]Bulk 1 is the bulk composition of the Jilin meteorite from Wang (1993) after subtraction of metal and troilite and recalculation to 100 %. Bulk 2 is the average bulk composition of areas free of metal and troilite in shock-induced silicate melt glass determined by the area analysis (EDS) technique. The compositions of olivine and pyroxene were analyzed by the EPMA technique. No- number of analyses; P- shock pressure, n.d.-not detected

revealed that the bulk silicate melt in veins (Table 8.2, bulk 2) has an identical composition to bulk Jilin chondrite (Table 8.2, bulk 1), indicating that a whole-rock melt was produced. At a shock pressure of 78 GPa, about fifteen percent of whole-rock melting occurred, whereas at 83 GPa we observed that twenty percent of the whole rock was melted. Most of the melt now occurs as an assemblage of a glass phase and very fine-grained crystallites of olivine due to subsequently fast cooling and quenching after impact. The presence of microcrystalline olivine and pyroxene in association with pyroxene glass and a glass mixed with a feldspathic phase in the Jilin samples shocked at 83 GPa was proved by micro-Raman spectroscopic studies (Fig. 8.2). Bubbles of 2–15 μm in diameter are observed in the shock-induced silicate melt.

At a peak pressure of 93 GPa, up to eighty percent of the whole rock was brecciated into grain sizes <20 μm. Forty percent of this fine-grained material has grain sizes <5 μm. It is common that brecciated olivine and pyroxene fragments occur in the plagioclase melt glass, indicating the melt was highly fluid at high pressures and temperatures. Up to thirty percent of the whole-rock melting was achieved. Although the majority of shock-induced silicate melt now occurs as a vesicular glass phase in the most heavily shocked samples, we still found evidence of crystallization of the melt in this sample. Figure 8.1c shows a melt pocket that consists of quenched dendritic microcrystalline olivine and pyroxene in grain sizes 1–3 μm, metal, troilite, and silicate melt glass. The fine-grained olivine and pyroxene were liquidus phases which crystallized from a shock-induced silicate

Fig. 8.2 Raman spectra of
minerals and silicate melt in
shock-loaded Jilin at 83 GPa.
a and **b** are from residual
olivine in silicate melt; **c–
e** are from silicate melt with
microcrystalline olivine; **f** is
from silicate melt with both
microcrystalline olivine and
pyroxene; **g** and **h** are from
silicate melt with
microcrystalline pyroxene
(Xie et al. 2001a)

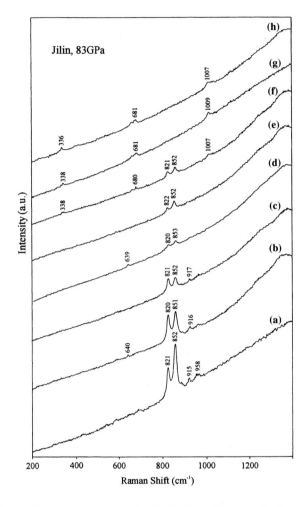

melt. Raman spectra of olivine fragments in shock-loaded Jilin at 93 GPa display
two strong peaks at 852 and 821 cm^{-1}. The Raman spectra of silicate melt also
display two weak bands at 847–853 cm^{-1} and 821–822 cm^{-1}, indicating that
shock-induced olivine normative melt does not quench into glass phase upon
cooling, but crystallized into fine-grained olivine crystals.

Samples shock-loaded at 133 GPa display pervasive melting that reached 60 % or
more by volume. Shocked samples usually consist of a mixture of heavily brecciated
silicate fragments and silicate melt in interstices, or silicate melt pockets with some
silicate fragments (Fig. 8.1d). Our micro-Raman spectroscopic studies revealed the
presence of microcrystalline olivine and pyroxene in this shock-induced melt.
Vesicles are commonly observed in shock-induced melt regions.

It is important to point out that no solid-state transformation of olivine and
pyroxene into high-pressure polymorphs, no dissociation reactions of olivine to
magnesiowüstite + majorite (or + perovskite), and no other high-pressure phases

were encountered in all Jilin samples that experienced pressure from 12 to 133 GPa. Vesiculation of the silicate melt is a common phenomenon in Jilin samples shocked at pressures above 78 GPa.

The chemical composition of minerals shocked at 12, 53, and 83 GPa was analyzed by EPMA. From Table 8.2, we can see that with the increasing shock pressure, the chemical composition of olivine and pyroxene remains unchanged.

8.2.3 Shock Effects in Naturally Shocked H- and L-Group Meteorites

It is known that shock deformation features, such as undulatory extinction, mosaicism, planar fractures, and planar deformation features in olivine and diaplectic glass of plagioclase, are similar in naturally shocked H and L chondrites, and they are commonly used for pressure calibration for weakly to moderately shocked meteorites (shock stages 2–4). The only difference was found in the mineral constituents of the shock melt regions in these two group chondrites (Xie et al. 1991, 2001c).

As we know that many of H-group chondrite meteorites are pervasively intersected by shock melt veins (up to 15 mm in thickness) and contain large melt pockets (few cm across) (Xie et al. 1991). A detailed study of the melt regions in some H-chondrites revealed that the melt veins and pockets consist mainly of two lithologies: (1) fine-grained glassy matrix with dendritic intergrowth of quenched olivine and pyroxene (below $\sim$5 μm) and blebs of FeNi and troilite in eutectic intergrowth; (2) coarse-grained rounded to irregular silicate grains of 10–200 μm in size (Xie et al. 1991, 2001a, b; Chen et al. 1995a, b). Kimura et al. (1999, 2000) conducted mineralogical study of heavily shocked Antarctic H-chondrites, Y-75277, Y-790467, Y-791524, Y-791949, Y-793535, Y-793537, ALH-78108, and A-880993, and revealed that all these H-group chondrites consist of highly deformed chondritic hosts and shock-induced melt regions. Raman spectroscopic investigations indicate that the melt regions mainly comprise coarse-grained and fine-grained low-pressure silicate minerals, and no high-pressure polymorphs of silicate phases were observed in any lithologies.

The heavily shocked Yanzhuang H6 chondrite displays shock deformations in different shock levels. Chen and Xie (1996) observed the presence of four distinct shock-metamorphosed regions with distinct deformations (Fig. 8.3a): (1) shocked chondritic region (40 % by volume of meteorite) with several sets of planar fractures in olivine (Fig. 8.3b) and weak mosaicism in silicate minerals; (2) brecciated chondritic region (20 % by volume of meteorite) with silicate fragment sizes <15 μm and strong mosaicism in olivine and pyroxene, and plagioclase glass; (3) partially melted, recrystallized, and brecciated chondritic region (10 % by volume of meteorite) with quenched plagioclase melt glass in the direct vicinity of shock melt veins; and (4) shock melt veins (up to 15 mm in width) and large melt

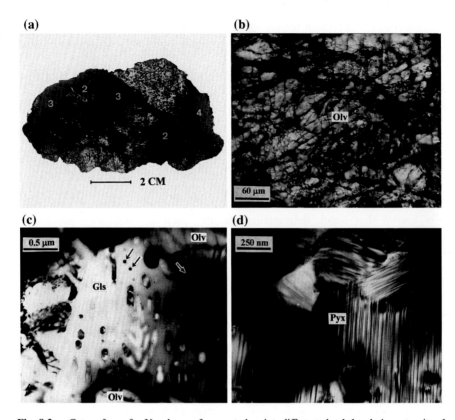

Fig. 8.3 a Cut surface of a Yanzhuang fragment showing different shock levels in meteorite: *1* shocked chondritic region; *2* brecciated chondritic region and partially melted, recrystallized, and brecciated region; *3* shock melt veins, and *4* large melt pocket. **b** Multiset planar fractures in olivine (*Olv*) in the chondritic region of Yanzhuang. Plane polarized light. **c** TEM bright field image showing quenched glass phase with idiomorphic and dendritic microcrystalline olivine in Yanzhuang shock melt vein. Note the tiny round vesicles (*arrows*) in quenched glass. Gls, silicate melt glass in interstice of silicates; Olv, olivine. **d** TEM bright field image shows the recrystallized polycrystalline pyroxene (*Pyx*) (Xie et al. 2001a)

pockets (up to 30 mm in size and 30 % by volume of meteorite) consisting mainly of a quench glass with idiomorphic or dendritic microcrystalline olivine and pyroxene with grain sizes <5 μm (Figs. 8.3c, d), recrystallized fine-grained poly-crystalline olivine and pyroxene, and stretched FeNi–troilite eutectic nodules. Some coarse-grained olivine and pyroxene fragments in the size of 100–200 μm can also be observed as remnants in the melt regions. These fragments were heavily shock-damaged and display very low birefringence under microscope. Although the melt regions in Yanzhuang mainly consist of low-pressure silicate minerals, high-pressure mineral phases, such as ringwoodite and majorite, were discovered in the Yanzhuang silicate melt regions (Chen and Xie 1993a, b). Two occurrences of ringwoodite were observed. One occurs in the form of mineral inclusions of 1–2 μm in size in the large olivine fragments in association with olivine diaplectic glass and recrystallized olivine, and another is as idiomorphic grains of 0.1–1.5 µm in size emerged in shock-induced silicate melt. Majorite in Yanzhuang occurs in the form of subhedral or subrounded grains of <1 µm in size in shock-damaged large pyroxene fragments as remnant high-pressure phase. Both ringwoodite and majorite were identified by Raman spectrometry. Ringwoodite displays a strong peak at 796 cm^{-1}, and majorite shows three main peaks at 926, 660, and 590 cm^{-1} on the Raman spectra of host olivine and pyroxene fragments, respectively (Chen and Xie 1993a, b). Another good example is the Antarctic H-chondrite Y-75100, which contains abundant high-pressure polymorphs of silicate minerals, such as ringwoodite, wadsleyite (β-spinel phase of olivine), majorite, jadeite, and hollandite-structured plagioclase, which were identified by micro-Raman investigations in its shock melt veins (Kimura et al. 1999, 2000). Tomioka and Kimura (2003) reported a shock-produced mixture of fine-grained majorite and $CaSiO_3$-rich glass transformed from diopside in an H chondrite. Therefore, we consider that the shock melt regions in some of the H-chondrites may also contain high-pressure polymorphs of silicate minerals.

The heavily shocked L-group chondrites contain unmelted but shock-deformed chondritic rock and shock melt veins. The shock melt veins in L-group chondrites are full of high-pressure phases. For example, the shock melt veins in the Sixiangkou L6 chondrite contain two distinct types of high-pressure assemblages (Chen et al. 1996): (1) multimineralic coarse-grained fragments consisting of the high-pressure polymorphs of olivine, ringwoodite and wadsleyite, and the high-pressure polymorphs of orthopyroxene, majorite, that were formed by solid-state phase transformation during shock compression (Chen et al. 1996a) and (2) a fine-grained high-pressure assemblage consisting of majorite–pyrope$_{ss}$ + magnesiowüstite. This assemblage is inferred to have crystallized from a dense shock-induced chondritic melt as liquidus phases upon cooling under high pressure. Variations in the composition of the liquidus assemblages in various chondrites were found, e.g., majorite–pyrope$_{ss}$ + magnetite in Tenham (Sharp et al. 1997a, b) and majorite–pyrope$_{ss}$ + ringwoodite + wadsleyite in Peace River and Mbale (Chen et al. 1996b, 1998b). These differences reflect differences in peak shock pressures in the melt veins of these chondrites. Not long ago, the high-pressure polymorph of plagioclase, hollandite, was also found as a liquidus phase in shock melt veins in several L-chondrites (Gillet et al. 2000; Tomioka and Fujino 1999; Xie et al. 2001b). All shock melt veins consisting of the high-pressure polymorphs do not show any sign of vesiculation, thus unequivocally indicating that these veins solidified under high pressure. According to the results of high-pressure melting experiments on Allende carbonaceous chondrite (Agee et al. 1995) and on peridotite (Zhang and Herzberg 1994), the presence of ringwoodite constrains the pressure to be greater than about 20 GPa and the majorite–pyrope garnet + magnesiowüstite assemblage crystallized from 2050 to 2300 °C and 20–24 GPa.

Although there are some differences in the mineral constituents of shock melt veins and pockets in H- and L -type of chondrites reflecting the difference in their shock history, there exist some similarities in shock *P-T* conditions for these two groups, since chondrites of both groups contain melt veins and melt pockets implying that their parent bodies experienced similar heavy collision events in the space. The discovery of ringwoodite, majorite, hollandite, and other high-pressure polymorphs of silicate minerals in shock-melt regions in some H-group chondrites indicates that phase transformation of silicate minerals did occur upon shock. Therefore, the *P-T* conditions for both L- and H-types of chondrites should also be similar. The only difference between them is in the post-shock temperatures and cooling rates. Considering the fact that the width and volume of shock melt veins and pockets in H-group chondrites are much larger than those for L-group chondrites, we assume that post-shock temperature in melt regions in H-group chondrites must be higher and lasted for longer time in comparison with that for L-group chondrites, thus leading to the retrograde metamorphism of most of the shock-formed high-pressure phases to their low-pressure phases, whereas the shock-formed high-pressure phases in the rather thin shock melt veins in L-group chondrites were survived for the high cooling rate after pressure release.

8.2.4 Comparison of Experimentally and Naturally Shocked Chondrites

The results of our shock-loading experiments on the Jilin H5 chondrite up to 53 GPa are comparable to shock experiments conducted to peak pressures of up to 60 GPa (Schmitt and Stöffler 1995) and to those conducted to peak pressures of ~ 40 GPa (Sears et al. 1984) on the H6 chondrite Kernouve. They are all almost exclusively characterized by mechanical disaggregation. Shock experiments contacted on olivine or dunite (with no FeNi metal and troilite in mineral composition) have not produced any high-pressure polymorphs of olivine or pyroxene (Reimold and Stöffler 1978). In addition, shock-loading experiments up to 70 GPa on the Leedy L6 chondrite failed to produce any high-pressure phase or whole-rock melting (Bogard et al. 1987). The results of our experiments on Jilin H5 chondrite show that the chondritic melts were firstly obtained only at P > 78 GPa, and more than 60 % whole-rock melting was achieved at P $\sim$ 133 GPa, but we also failed to produce any high-pressure phases in Jilin samples shock-loaded at pressure range of 12–133 GPa. It is evident that the influence of the bulk chemistry and mineralogical composition of the chondrites on the results of shock experiments is minor.

Shock-induced deformation features in the experimentally shocked Jilin samples resemble those observed in some different shock-metamorphosed regions in the naturally shocked Yanzhuang H-chondrite and Sixiangkou L-chondrites. Namely, the shock features of planar fractures in olivine, mosaicism in olivine and pyroxene, and transformation of plagioclase into glass phase in the Jilin samples shocked at 27, 39, and 57 GPa are similar to those observed in the regions (1) and (2) in Yanzhuang meteorite and to the unmelted chondritic rock in the Sixiangkou L-chondrites, and some shock features in Jilin shocked up to 93 and 133 GPa, such as brecciation, whole-rock melting can be compared with those found in shock regions (3) and (4) in Yanzhuang and in the shock melt veins in Sixiangkou L-chondrite. However, the shock features in Jilin samples shock-loaded at pressures higher than 78 GPa are distinct from those encountered in heavily shocked Yanzhuang and Sixiangkou type of chondrites (Chen and Xie 1997; Chen et al. 1996a; Stöffler et al. 1991; Sharp et al. 1997a, b; Xie et al. 2001a). The main differences are discovered in the mineral assemblages in the melt regions. The shock-induced melt regions in the Jilin samples that experienced pressures from 78 to 133 GPa contain only low-pressure mineral assemblages, and no solid-state transformation of olivine and pyroxene into high-pressure polymorphs, no dissociation reactions of olivine to magnesiowüstite + majorite (or +perovskite), and no other high-pressure phases were encountered. On the other hand, the crystallized melts in naturally shocked chondrites contain different high-pressure phases.

The difference in the type of mineral assemblages in shock-melt regions in experimentally and naturally shocked chondrites was not noticed before (Stöffler et al. 1991; Xie et al. 1991). However, the presence of high-pressure assemblages on the one hand, and low-pressure assemblages on the other hand associated with shock-induced melting indicates distinct differences in the P-T conditions and the duration of the shock regimes on chondritic asteroids.

Chen et al. (1996a, 1998a) proposed a new hypothesis in that the differences in low-pressure mineral and high-pressure mineral assemblages in melt veins of experimentally and naturally shocked chondrites directly related to duration of the high-pressure regime and the cooling history of the veins. The presence of coarse-grained high-pressure phases, especially the presence of fine-grained liquidus garnet + magnesiowüstite in melt veins of Sixiangkou L-chondrites, and that of high-pressure phases in the Yanzhuang H-chondrites, as well as the absence of vesicles in melt regions of both type chondrites, implies that all these minerals crystallized under high pressure and high temperature, which were estimated to be in the range of 20–24 GPa, and in excess of 2000 °C (Chen et al. 1996a), and indicates that the duration of high-pressure regime in shock veins should be long enough (several seconds) for the phase transformation of coarse-grained minerals in solid state and for the crystallization of fine-grained high-pressure matrix phases from silicate melt. The small grain size of the liquidus phases (a few micrometers) implies that the cooling rate of the vein materials was high so that all high-pressure phases were preserved in the veins. On the other hand, the duration of high-pressure regime for the shocked Jilin samples was very short, because most of the shock phenomena during shock experiments using the gun method are completed within a microsecond timescale (Goto and Syono 1984), although the shock-induced peak pressures and shock temperatures in experimentally shocked Jilin samples were very high (up to 133 GPa and 2400 °C) that caused pervasive whole-rock melting of the chondritic rock. Evidently, such regime is too short to complete the formation of high-pressure phases in veins. The sign of vesiculation in the shock-induced melt in Jilin samples indicates that the duration of high-pressure regime was short and the solidification and crystallization of shock-induced melt took place after pressure release.

Although the shock stages of Stöffler et al. (1991) correctly reflect the sequence of increasingly more metamorphosed material, the corresponding pressure calibration provides very high shock pressure estimates. The pressure calibration of Stöffler et al. (1991), which is based on shock recovery experiments, defines the S1/S2, S2/3, S3/S4, S4/S5, and S5/S6 transitions as <5 GPa, 5–10 GPa, 15–20 GPa, 30–35 GPa, and 45–55 GPa, respectively, with S6 conditions including pressures up to 90 GPa. Sharp and DeCarli (2006) correctly summarized the reasons why such pressure estimates are too high. They pointed out that although ordinary chondrites are porous, the samples used in most shock recovery experiments, which commonly include single crystals and igneous rock fragments, are not porous (Schmitt 2000). According to the porous Hugoniots, the internal energy increase of compression in these nonporous samples is much higher for initially porous materials. Recognizing that shock recovery experiments on nonporous samples result in relatively low shock temperatures, Schmitt (2000) did shock recovery experiments at elevated starting temperatures (920 K) as well as at low temperature (293 K) to investigate the temperature effects on shock metamorphism pressure for the H6 chondrite Kernouve. The importance of kinetics in shock metamorphism is illustrated by the fact that maskelynite formation occurred at 20–25 GPa in the preheated experiments compared to 25–30 GPa in the low-temperature samples and 30–35 GPa in the Stöffler et al. (1991) study. These and other high-temperature shock experiments demonstrate that shock-metamorphic effects are temperature dependent as well as pressure dependent and that one cannot accurately calibrate the shock pressures without kinetic effects. If kinetic effects are indeed important, one may further question the relevance of μs duration shock recovery experiments to much longer duration natural shock events. In addition to the porosity problem, Sharp and DeCarli (2006) pointed out that shock recovery experiments done in high shock impedance sample containers result in high shock pressures and relatively low shock temperatures (Bowden 2002). According to the rules on shock reflections and loading path effects, samples in high-impedance containers reach peak (continuum) pressure via a series of shock reflections, with the result that shock and post-shock temperature are substantially lower than for a sample loaded via a single shock to the same peak pressure (Bowden 2002; DeCarli et al. 2002). For example, Bowden (2002), who shocked quartz sand in containers of various shock impedances, produced multiple intersecting planar features at 8 GPa in impedance-matched polyethylene containers and at 19 GPa in high-impedance stainless steel containers. Since most of the shock recovery data in the Stöffler et al. (1991) calibration were from experiments in high-impedance containers, the calibrated pressures of thermally activated shock effects, such as phase transformations, are likely to be too high. Therefore, we consider that the textural evidence for phase transformations of silicate minerals in melt veins in many L-group and some H-group chondrites indicate that the shock pulses that caused the transformations in natural samples were of relatively long duration, perhaps up to several seconds (Chen et al. 1996a, 2004a; Xie et al. 2001c; Sharp and DeCarli 2006). The microsecond duration typical of shock experiments, combined with the relatively low shock temperatures produced in shock recovery experiments, can very well explain why such experiments have not produced high-pressure polymorphs of silicate minerals. The kinetics of such reconstructive phase transformations are simply too low under conditions of shock recovery experiments.

Sharp and DeCarli (2006) point out that phase transformations occurring in meteorites are strongly dependent on temperature as well pressure, because reconstructive phase transitions in silicates are kinetically sluggish and require high temperatures to overcome the large activation barriers. The fact that olivine and pyroxene do not transform at very high pressures in diamond anvil experiments without being heated to very high temperatures further supports the idea that high temperature actually controls the distribution of olivine and pyroxene high-pressure polymorphs in chondrites. The shock calibration of Stöffler et al. (1991) stresses shock pressure as primary driver of shock-metamorphic effects and does not discuss temperature effects or reaction kinetics. The static high-pressure kinetic experiments have shown that dry hot-pressed San Carlos olivine transforms to ringwoodite, on an observable timescale, only above 900 °C at 18–20 GPa (Kerschhofer et al. 1996, 1998, 2000). The transformation of enstatite to akimotoite is even more sluggish, requiring temperatures in excess of 1550 °C for transformation at 22 Gpa (Hogrefe et al. 1994). The fact that some olivine and pyroxene in meteorites transform to high-pressure polymorphs by nucleation and growth indicates that shock temperature of the transformed material must have been much higher than the temperatures in the static experiments. This is supported by the observation that solid-state transformations of olivine and pyroxene occur almost exclusively within or in close proximity to shock front (Kieffer 1971). Sharp and DeCarli (2006) assumed that the solid-state transformations that occur in and along melt veins in meteorites are likely to have formed at the equilibrated shock pressure rather than during transient pressure spikes because the elevated temperature of the melt vein regions can last up to about second.

8.2.5 Feasibility of Using Static High-Pressure Melting Experiments to Estimate the P-T Conditions of Natural Shock Events

As we introduced in the previous section that the pressure calibration of Stöffler et al. (1991), which is based on shock recovery experiments, gives too high-pressure estimates for naturally shocked meteorites. Chen et al. (1996a) indicated that the duration of shock-induced high-pressure regime is of great significance in the formation of high-pressure phases in meteorites. They point out that it may appear impossible to retain high pressure and temperature for up to several seconds on the basis of shock experiments, but one can envisage that collision of large asteroidal bodies or the passage of multiple shock waves through such bodies during complex collisional events could account for such conditions. This implies that the high-pressure and high-temperature conditions in melt veins of naturally shocked meteorites could last for several seconds, and the results of long-duration natural shock events can approximately be compared with those of static high-pressure experiments. It is known that the olivine–ringwoodite, enstatite–majorite, enstatite–akimotoite, and enstatite–perovskite phase transitions are all metastable, and the metastable phase boundaries of interest are lower in pressure than the minimum pressure of stable equilibrium. The phase relation only limits the pressure to be greater than the metastable equilibrium between the low- and high-pressure polymorphs. Ringwoodite that occurs in or adjacent to shock melt in heavily shocked chondrites formed at temperatures much higher than those in shock recovery experiments and over a shock duration that may exceed the experimental shock duration by five orders of magnitude (Sharp and DeCarli 2006). Ringwoodite in shocked chondrites must form at pressures in excess of the metastable phase boundary (～ 18 GPa), but the amount of pressure overstepping required in nature is not constrained by shock or static high-pressure experiments. An alternative means of constraining shock pressure in meteorites using solid-state transformation effects are to use the crystallization of high-pressure minerals from the shock-induced melt combined with experimental high-pressure melting relations. This approach was first used by Chen et al. (1996a), who used TEM to determine melt-vein assemblage in Sixiangkou chondrite. They found the chondritic melt in a large melt vein crystallized to form majoritic garnet and magnesiuowüstite, which based on the phase diagram of the CV3 chondrite Allende (Agee et al. 1995) is stable at pressures between about 22 and 27 GPa (Fig. 8.4). Chen et al. (1996a) inferred a

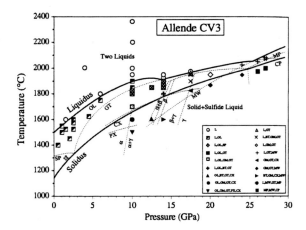

Fig. 8.4 Pressure versus temperature phase diagram for Allende meteorite (After Agee 1995). *Bold lines* are silicate liquidus and solidus. *Lighter dashed lines* ate the major phase boundaries identified in Allende experiments. "Two liquidus" denotes the field of superliquidus immiscibility between silicate and sulfide. L = silicate liquid; OL = olivine, SP = Cr–Al–Mg spinel; GT = garnet; GM = gamma spinel; BT = beta phase; CX = diopsidic clinopyroxene; PX = low-Ca clinopyroxene; MW = magnesiowüstite; MP = (Mg,Fe)SiO$_3$ perovskite. CP = CaSiO$_3$ perovskite

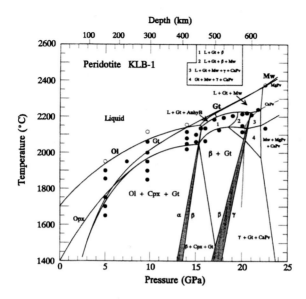

Fig. 8.5 Phase diagram for anhydrous peridotite KLB-1 (After Zhang and Herzberg 1994). All crystalline (*solid circles*), crystals plus liquid (*cross-hatched circles*), all liquid (*open circles*), and precise liquidus determination at 2375 and 22.5 GPa (half open and half *cross-hatched circles*). L = liquid; Ol = olivine; Opx = orthopyroxene; Cpx = clinopyroxene; Gt = garnet; Anhy B = anhydrous B; Mw = magnesiowüstite; MgPv = magnesium silicate perovskite; CaPv = calcium silicate perovskite; α = olivine; β = modified spinel $(Mg,Fe)_2SiO_4$; and γ-spinel $(Mg, Fe)_2SiO_4$

crystallization pressure of about 25 GPa, which is half the value of the low-pressure threshold for S6 shock conditions of Stöffler et al. (1991). Sharp and DeCarli (2006) indicate that there are several reasons why high-pressure melting relations can be applied to the interpretation of melt-vein crystallization. First, the most common melt-vein assemblage seen in S6 chondrites, majorite plus magnesiuowüstite, is also produced in static high-pressure melting experiments on the Allende (Agee et al. 1995) and on Kilburn-hole-1 peridotite (Zhang and Herzberg 1994) (Fig. 8.5). The textures and crystal sizes in the centers of large chondritic melt veins, such as those in Sixiangkou, Tenham, and Acfer 040 (Chen et al. 1996a; Sharp et al. 1997a, b; Aramovich et al. 2003), are very similar to the textures and crystal sizes produced in the static experiments (Agee et al. 1995). Similarly, the chemical compositions of the crystallized majoritic garnets are very similar to the compositions of garnets in the experiments (Chen et al. 1996a). Compared to solid-state reconstructive phase transitions, melt-vein crystallization involves much smaller kinetic barriers. Hence, it is plausible to use the static melting experiments to make more accurate estimates for *P-T* histories of shocked meteorites.

8.3   First Evaluation of *P-T* History of the Suizhou Meteorite

As we described in Chap. 7, the shock effects in olivine and pyroxene in the Suizhou meteorite indicate that these two rock-forming minerals were shocked to stage S4 or even to S5. Most plagioclase grains in the Suizhou unmelted chondritic rock was melted and transformed into maskelynite during the shock event. This implies that the Suizhou meteorite should have strongly shocked to stage 5. The presence of shock melt veins and transition of olivine into ringwoodite indicate that these parts of the meteorite were very strongly shocked to stage 6. Based on the classification and pressure calibration of shock stages defined by Stöffler et al. (1991), the whole rock of the Suizhou meteorite could be strongly shocked up to stage S5 in the pressure range of 35–45 GPa, with some localized regions (veins) shocked to stage 6 in the pressure range of 45–90 GPa. Hence, the estimated shock pressure for this meteorite could be in the range of 35–90 GPa.

8.4   Actual *P-T* History of the Suizhou Meteorite

According to the scheme defined by Stöffler et al. (1991) on the basis of shock experiments, the shock-produced mineralogical features in the Suizhou meteorite match shock stages from S5 to S6 and cover a wide range of high pressures from 35 to 90 GPa. Such pressure estimation is too high and is not consistent with the results of static high-pressure experiments on Allende chondrite. Therefore, we must to find out the reason that causes such an unusual estimation and evaluate the actual *P-T* history of the Suizhou meteorite.

8.4.1   *P-T History of Suizhou Meteorite Evaluated from Melting of Plagioclase*

Shock wave experiments have shown that feldspar transforms to diaplectic glass at 26–34 GPa and to melt glass at 42 GPa (Ostertag 1983). It is known that an experimental shock pressure of 30 GPa at the ambient temperature of the chondritic rocks will only produce a temperature rise of less than 350 °C (Stöffler et al. 1991). Such a temperature is too low to melt plagioclase although the shock pressure is very high. However, the transition process may be accelerated at pressures as low as to 20–25 GPa when the temperature is high enough (Schmitt 2000). The fact that most plagioclase grains in the Suizhou meteorite had transformed into maskelynite indicates that the shock-produced temperature must be far higher than 350 °C. The investigation on the basaltic meteorite Zagami by Langenhorst and Poirier (2000) revealed that the shock pressure and equilibrium shock temperature might reach

30 GPa and 1000 °C, respectively, since all plagioclases in this meteorite were transformed into maskelynite (Langenhorst et al. 1991; McCoy et al. 1992; Chen and El Goresy 2000).

It has been emphasized recently by Chen and El Goresy (2000) that phase transitions occurring in many natural impact events could correspond not only to dynamic but more likely to kinetic processes as well. The high-pressure pulse in the natural impact events might last for several seconds in large impact bodies (Chen et al. 1996a). The extended period of higher pressure and temperature regime in the natural impact events might play an important role in the phase transition of minerals in chondrites (Chen et al. 1996a; Sharp et al. 1997a, b; Tomioka and Fujino 1997; Gillet et al. 2000; Xie et al. 2001c). We assume that maskelynite could be formed at a shock pressure lower than that in experiments if the duration of high-pressure and temperature regime is long enough. Therefore, we estimate that the Suizhou meteorite might be shocked in a pressure range of 25–30 GPa and at a temperature of about 1000 °C.

8.4.2 P-T History of Suizhou Melt Veins Evaluated from High-Pressure Mineral Assemblages

The melt veins in the Suizhou meteorite contain abundant high-pressure phases, including coarse-grained ringwoodite, majorite, akimotoite, perovskite glass, linguite, and fine-grained majorite–pyrope$_{ss}$ + magnesiowüstite + ringwoodite. According to the experimental data of Katsura and Ito (1989), the formation of ringwoodite with about 80 mol% of Mg_2SiO_4 constrains the pressure to ~ 18 GPa at 1600 °C. The existence of linguite instead of calcium ferrite-type $NaAlSiO_4$ constrains the pressure not higher than 23 GPa at 1000–1200 °C (Liu 1978; Yagi et al. 1994). Experimental results on the Allende meteorite by Agee et al. (1995) shows that the assemblage majorite garnet + ringwoodite stabilizes under *P-T* conditions of 18–22 GPa and 1800–1900 °C and the assemblage majorite–pyrope$_{ss}$ + magnesiowüstite is stable at pressures between 22 and 27 GPa and at temperatures between 2000 and 2400 °C. Therefore, the occurrence of abundant high-pressure phases including ringwoodite, majorite, akimotoite, perovskite, linguite, and majorite–pyrope$_{ss}$ + magnesiowüstite in the shock melt veins of Suizhou constrains the pressure and temperature to be 18–27 GPa and 1800–2400 °C, respectively.

Shock compression experiments so far have revealed that most silicates and oxides undergo shock-induced phase transitions in the pressure range up to 200 GPa (Ahrens et al. 1969; Ahrens 1980). The phase transition induced by a shock wave can be classified into two major categories, reconstructive type and non reconstructive type, depending upon whether it requires atomic diffusion or not. The phase transition of the nonreconstructive type may be completed within a microsecond timescale of shock compression experiments. Martensitic or electronic transition belongs to this category. However, if the phase transition is of reconstructive nature in which time sufficient for atomic diffusion is required to complete the transition, the situation is no longer simple. A substantial amount of pressure overdriving is necessary for the onset of the phase transition, and the wide mixed phase region is observed before entering the high-pressure phase region. The phase transition occurred in complex silicates such as olivine and pyroxene is essentially of the reconstructive type. As olivine is the most important constituent in the Earth's mantle, shock compression studies were extensively carried out on polycrystalline olivine and olivine rocks. Indication of the phase transition was noted above 40 GPa, which is substantially higher than the equilibrium phase transition pressure observed in static high-pressure experiments (Syono 1984). For example, a comprehensive study of shock compression measurements of single crystal of forsterite (Mg_2SiO_4) was performed up to 170 GPa (Jackson and Ahrens 1979; Syono et al. 1981). The onset of the phase transition was clearly demonstrated at about 50 GPa, which is remarkably higher than that of known high-pressure transformations in static experiments, e.g., below 20 GPa for olivine-modified spinel or spinel transition (Akimoto et al. 1976) and below 30 GPa for post-spinel dissociation to ilmenite or perovskite-type $MgSiO_3$ plus MgO (Ito et al. 1984), and also slightly higher than that observed in shock compression of polycrystalline olivines.

As we noted previously, ringwoodite, the high-pressure phase of olivine, has not been found in shock recovery experiments carried out above 56 GPa (Jeanloz et al. 1977), and it has not been observed in all chondrites with localized melting and recrystallized olivine. Steel and Smith (1978) assumed that special conditions, e.g., elevated temperature of the target before shock compression, may be required for the formation of ringwoodite. We argue that the extended duration (several seconds) of high-pressure and high-temperature regime in large naturally impacted bodies may provide such favorable temperature conditions for phase transition of minerals when these impacted bodies are still at high pressure and the shock-induced high temperature in such bodies may reduce the pressure range required for phase transition of minerals. If this is the case, it would be reasonable to reduce the upper limit of high-pressure range from 27 GPa to 25 GPa for the Suizhou melt veins. So we set the pressure range for the Suizhou melt veins at 18–25 GPa.

The above result indicates that minerals in shock melt veins in the Suizhou meteorite were shocked to pressures similar to those in the unmelted host chondritic rock (20–25 GPa), but to temperatures much higher than those in its unmelted host rock ($\sim$ 1000 °C). Therefore, additional heat is needed to raise the temperature of materials in veins.

It has been revealed that the stress at the shock compression often concentrates along the shear zone, which is often close to the direction of maximum resolved stress, as schematically shown in Fig. 8.6, that demonstrates the temperature profile along the line *XY* is schematically shown in the lower part of the figure, where T_H is average Hugoniot temperature calculated from conservation relations, and T_S is the temperature expected for isentropic compression. The enlarged section in the circle in Fig. 8.6 demonstrates how the phase transition proceeds under shock compression in accordance with temperature increase. In this figure, A is high-temperature

Fig. 8.6 Schematic illustration of heterogeneous yielding in the shock-induced brittle substances with low thermal conductivity (after Syono 1984)

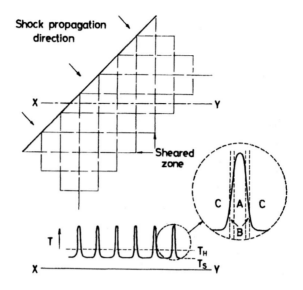

shear zone where melting may occur; B is transition zone where solid-state reconstructive phase transition is possible; and C is low temperature region in which materials are compressed only isentropically and remain relatively undamaged (Syono 1984). Hence, these sheared zones are considered to be brought to high temperatures, often exceeding the melting temperature, while the region between the sheared zones should be only isentropically compressed and its temperature remains low (Fig. 8.6). Such temperature distribution will not be equilibrated within a timescale of microsecond duration of shock pulse because of the low thermal conductivity of these materials (Syono 1984).

It is clear that when the high-pressure phase transition of the reconstructive type is involved, reactions will be preferentially occur along the sheared zone where temperature is high enough to promote the phase transition, resulting in extraordinarily heterogeneous textures consisting of high-pressure reaction products and a relatively undamaged portion in the shock-recovered materials.

In the case of melt-vein-bearing chondrites, the shock compression pressure for both unmelted and melted portions should be the same and can be evaluated by the shock-induced effects in olivine and plagioclase. However, the temperature for unmelted portion would much lower than that in melt veins. According to the theory of heterogeneous yielding in the shock-induced brittle substances with low thermal conductivity described above, additional heat might be induced in the sheared zones by shear friction process. The shock melt veins are considered to be the sheared zones, e.g., the zone A in Fig. 8.6, and the unmelted chondritic rock are considered to be the low-temperature region in which materials are compressed only isentropically and remain relatively undamaged, e.g., the zone C in Fig. 8.6.

As we mentioned before that the unmelted portion of the Suizhou chondrite experienced average shock pressure and temperature of about 22 GPa and 1000 °C on the basis of transformation of most plagioclase grains into maskelynite, and the locally developed shock veins in Suizhou were formed at pressures of 22–25 GPa that is close to the shock pressure experienced by the Suizhou unmelted chondritic rock but at an elevated temperature of about 1900–2000 °C (Xie et al. 2001a, b; Chen et al. 2004). Therefore, we must to answer the question of what was the additional heat source for the Suizhou shock veins. For this reason, we studied the morphological features of the Suizhou shock vein in more detail. It has been found that the main shock veins microscopically are oriented almost in the one direction and are paralleling with each other. Some diagonally intersected melt veins in between the two neighboring melt veins (Fig. 8.7), or in both sides of a single main vein (Fig. 8.8), can be observed. The angle between the straight and diagonal veins is about 45°. That is indicative of strong shearing movement along the veins during the shock compression. Besides, the sharp and straight boundaries between shock melt veins and the surrounding unmelted chondritic rock in the Suizhou meteorite are also indicative of strong shearing movement along the veins (Figs. 1.6, 3.1 and 3.2). These distinct morphological features of veins provide strong and direct evidence for a shearing origin of shock melt veins in the Suizhou meteorite. Hence, we assume that the higher temperature in melt veins than that in the unmelted Suizhou chondritic rock was achieved by localized shear stress excursions caused by the collision event. Hence, we finally set the *P-T* conditions for the Suizhou melt veins at 18–25 GPa and 1900–2000 °C.

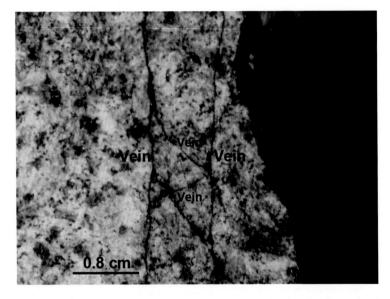

Fig. 8.7 Photograph of part of a Suizhou meteorite fragment showing the diagonal melt veins (*Vein*) in between two main parallel shock veins

Fig. 8.8 Photograph of part of a Suizhou meteorite fragment showing the diagonal melt veins in both sides of a main straight shock vein (*Vein*)

8.4.3 P-T History of the Suizhou Meteorite

As noted above, the host Suizhou chondritic rock experienced a shock-metamorphic regime of up to 25–30 GPa at $\sim$1000 °C, whereas the vein material was shocked at up to 18–25 GPa and 1900–2000 °C. We take 25 GPa as the upper limit of shock pressure for both the host chondritic rock and the veins.

The presence of small rounded FeNi metal grains deposited from the vapor phase in cracks and fractures joints in the olivine and pyroxene provides additional evidence of the shock temperature produced in the Suizhou meteorite. Experimental vaporization of an L6 chondrite conducted by Gooding and Muenow (1977) showed that the vapor pressure of Fe and Ni is very low ($<10^{-10}$ atm) at <800 °C, but it can reach $10^{-5.5}$ and $10^{-6.5}$ atm at 1200 °C. Therefore, these authors indicate that vaporization at <800 °C is not an efficient mechanism for significant transfer of a metal phase unless very long periods of time are available (Gooding and Muenow 1977). In fact, a temperature of 850–1300 °C was estimated for the formation of shock-produced FeNi metal deposited from the vapor phase in the thick-vein-bearing Yanzhuang chondrite (Xie and Chen 1997). Here, we estimate that the shock temperature in the Suizhou unmelted host chondritic rock may be about 1000 °C. This is in good agreement with the estimation for maskelynite in the chondritic rock.

Based on the shock features of minerals in both unmelted and melted domains, we consider that the Suizhou meteorite was strongly shock-metamorphosed, and the shock level of the meteorite would be stage 5. The shock pressure and temperature are estimated at 25 GPa and 1000 °C for the unmelted chondritic rock and 25 GPa and 2000 °C for the shock melt veins.

8.5   Summary

The Suizhou L6 chondrite contains moderately shocked olivine, pyroxene, metal, and troilite, but almost all plagioclase grains were melted and transformed into maskelynite during the shock event. A few very thin shock melt veins filled with abundant high-pressure phases are observed in the meteorite. The actual shock level of this meteorite was evaluated as stage S5. This meteorite experienced a shock pressure and temperature of up to 25 GPa and ~1000 °C, respectively. Locally developed thin shock veins in the meteorite were shocked to stage S6 and were formed at the same pressure (up to 25 GPa) but at an elevated temperature of about 2000 °C. The higher temperature in melt veins than in the unmelted chondritic rock was achieved by localized shear friction stress.

References

Agee CB, Li J, Shannon MC et al (1995) Pressure-temperature phase diagram for Allende meteorite. J Geophys Res 100:17725–17740

Ahrens TJ (1980) Dynamic compression of earth materials. Science 207:1035–1041

Ahrens TJ, Anderson DL, Ringwood AE (1969) Equation of state and crystal structures of high-pressure phases of shocked silicates and oxides. Rev Geophys 7:667–707

Akimoto S, Matsui Y, Syono Y (1976) High-pressure crystal chemistry of orthosilicates and the formation of the mantle transition zone. In: Streas RGJ (ed) Physics and chemistry of minerals and rocks. Wiley, London, pp 327–363

Aramovich CJ, Sharp TG, Wolf G (2003) The distribution and significance of shock-induced high-pressure minerals in chondrite skip Wilson. Lunar Planet Sci 34: abstract #1355.pdf

Begemann F, Palme H, Spettel B et al (1992) On the thermal history of heavily shocked Yanzhuang H-chondrite. Meteoritics 27:174–178

Bogard D, Hörz F, Johnson P (1987) Shock effects and argon loss in samples of the Leedy L6 chondrite experimentally shocked to 29–70 GPa pressures. Geochim Cosmochim Acta 51:2035–2044

Bowden KE (2002) Effects of loading path on the shock metamorphism of porous quartz: an experimental study. London: Ph.D thesis, University College, 228

Chen M, El Goresy A (2000) The nature of maskelynite in shocked meteorites: not diaplectic glass but a glass quenched from shock-induced dense melt at high-pressures. Earth Planet Sci Lett 179:489–502

Chen M, Xie XD (1993a) The shock effects of olivine in the Yanzhuang chondrite. Acta Mineralogica Sinica 13:109–114

Chen M, Xie XD (1993b) The shock effects of orthopyroxene in the heavily shocked meteorites. Chinese Sci Bull 38:1025–1027

Chen M, Xie XD (1996) Na behavior in shock-induced melt phase of the Yanzhuang (H6) chondrite. Eur J Min 8:325–333

Chen M, Xie XD (1997) Shock effects and history of the Yanzhuang meteorite: a case different from the L-chondrites. Chin Sci Bull 42:1889–1894

Chen M, Wopenka B, Xie XD et al (1995a) A new high-pressure polymorph of chlorapatite in the shocked chondrite Sixiangkou(L6). Lunar Planet Sci 26:237–238

Chen M, Xie XD, El Goresy A (1995b) Nonequilibrium solidification and micro- structures of metal phases in the shock-induced melt of the Yanzhuang (H6) chondrite. Meteoritics 30:28–32

Chen M, Sharp TG, El Goresy A et al (1996a) The majorite—pyrope + magnesiowustite assemblage: Constrains on the history of shock veins in chondrites. Science 271:1570–1573

Chen M, Wopenka B, El Goresy A (1996b) High-pressure assemblage in shock melt vein in Peace River (L6) chondrite: compositions and pressure-temperature history. Meteoritics 31(Suppl.): A27

Chen M, Xie XD, El Goresy A et al (1998a) Cooling rates in the shock veins of chondrites: constraints on the (Mg, Fe)$_2$SiO$_4$ polymorph transformations. Sci China Ser D 41:522–552

Chen M, Xie XD, El Goresy A (1998b) Olivine plus pyroxene assemblages in the shock veins of the Yanzhuang chondrite: constraints on the history of H-chondrites. Neus Jahrbuch für Mineral 3:97–110

Chen M, El Goresy A, Frost D et al (2004) Melting experiments of a chondritic meteorite between 16 and 25 Gpa: Implications for Na/K fractionation in a primitive chondritic Earth's mantle. Eur J Mineral 16:201–211

Dai C, Wang D, Jin X (1991) Shock-loading experimental study of Jilin meteorite. Chin Sci Bull 36:1984–1987

DeCarli PS, Bowden E, Jones AP et al (2002) Laboratory impact experiments versus natural impact events. In: Koeberl C, MacLeod KG (eds) Catastrophic events and mass extinctions: impacts and beyond. Boulder: Geological Society of America Special Paper 356, pp 595–605

Gillet P, Chen M, Dubrovinsky L et al (2000) Natural NaAlSi$_3$O$_8$ –hollandite in the Sixiangkou Meteorite. Science 287:1633–1637

Gooding JL, Muenow DW (1977) Experimental vaporization of the holbrook chondrite. Meteoritics 12:401–408

Goto T, Syono Y (1984) Technical aspect of shock compression experiments using the gun method. In: Sunagawa I (ed) Materials science of the Earth's interior. Terra Scientific Publishing Company, Tokyo, pp 605–619

Hogrefe A, Rubie DC, Sharp TG et al (1994) Metastability of enstatite in deep subducting lithosphere. Nature 372:351–353

Ito E, Takahashi E, Matsui Y (1984) The mineralogy and chemistry of the lower mantle: an implication of the ultrahigh-pressure phase relation in the system of MgO-FeO-SiO$_2$. Earth Planet Sci Lett 67:238–248

Jackson I, Ahrens TJ (1979) Shock-wave compression of single crystal forsterite. Geophys Res 84:3039–3048

Jeanloz R, Ahrens TJ, Lalley JS et al (1977) Shock produced olivine glass: first observation. Science 197:457–459

Jing F (1986) Introduction to experimental state equations. Science Press, Beijing, pp 172–173

Katsura T, Ito E (1989) The system Mg$_2$SiO$_4$-Fe$_2$SiO$_4$ at high-pressures and temperatures: precise determination of stabilities of olivine, modified spinel, and spinel. J Geophys Res 94:15663–15670

Kerschofer L, Sharp TG, Rubie DC (1996) Intracrystalline trans formation of olivine to wadsleyite and ringwoodite under subduction zone conditions. Science 274:79–81

Kerschofer L, Dupas C, Liu M et al (1998) Polymorphic transformation between olivine, wadsleyite and ringwoodite: mechanism of intracrystalline nucleation and the role of elastic strain. Mineral Mag 62:617–638

Kerschofer L, Rubie DC, Sharp TG et al (2000) Kinetics of intracrystalline olivine-ringwoodite transformation. Phys Earth Planet Inter 121:59–76

Kieffer SW (1971) Shock metamorphism of the Coconino sandstone at Meteor Crater, Arizona. J Geophys Res 76:5449–5473

Kimura M, El Goresy A, Suzuki A et al (1999) Heavily shocked antarctic H-chondrites: petrology and shock history. Antarct Meteorites 24:67–68

Kimura M, Suzuki A, Kondo T et al (2000) The first discovery of high-pressure polimorphs, jadeite, hollandite, wadsleyite and majorite, from an H-chondrite, Y-75100. Antarct Meteorites 25:41–42

Kondo K, Sawaoka A, Saito S (1977) Magnetoflyer method for measuring gas-gun projectile velocities. Rev Sci Instrum 48:1581–1582

Langenhorst F, Poirier JP (2000) 'Ecologitic' minerals in a shocked basaltic meteorite. Earth Planet Sci Lett 176:259–265

Langenhorst F, Stöffler D, Keil K (1991) Shock metamorphism of the Zagami achondrite. Lunar Planet Sci 22:779–780

Liu LG (1978) High-pressure phase transformations of albite, jadeite and nepheline. Earth Planet Sci Lett 37:438–444

McCoy TJ, Taylor GJ, Keil K (1992) Zagami: product of two-stage magmatic history. Geochim Cosmochim Acta 56:3571–3582

McQueen RG, Marsh SP, Taylor JW et al (1970) The equation of state of solids from shock wave studies. In: Kinslow R (ed) High-velocity impact phenomena. Academic Press, New York, pp 530–568

Ostertag R (1983) Shock experiments on feldspar crystals. J Geophys Res 88:B364–B376

Reimold WU, Stöffler D (1978) Experimental shock metamorphism of dunite. In: Proceedings of 9th Lunar Science Conference. pp 2805–2824

Schmitt RT (2000) Shock experiments with the H6 chondrite Kernouve: pressure calibration of microscopic shock effects. Meteorol Planet Sci 35:545–560

Schmitt RT, Stöffler D (1995) Experimental data in support of the 1991 shock classification of chondrites (Abstract). Meteoritics 30:574–575

Sears DW, Ashwort JR, Broadbent CP (1984) Studies of an artificially shock-loaded H group chondrite. Geochim Cosmochim Acta 48:343–360

Sharp TG, DeCarli PS (2006) Shock effects in Meteorites. In: Binzel RP (ed) Meteorites and the early solar system II. The University of Arizona Press, Tucson, pp 653–677

Sharp TG, Chen M, El Goresy A (1997a) Mineralogy and microstructures of shock-induced melt veins in the Tenham (L6) chondrite. Lunar Planet Sci 28:1283–1284

Sharp TG, Lingemann CM, Dupas C et al (1997b) Natural occurrence of $MgSiO_3$-ilmenite and evidence for $MgSiO_3$–perovskite in a shocked L chondrite. Science 277:352–355

Steel IM, Smith JV (1978) Coorara and Coolamon meteorites: ringwoodite and mineralogical differences. Lunar Planet Sci 9:1101–1103

Stöffler D, Keil K, Scott ED (1991) Shock metamorphism of ordinary chondrites. Geochim Cosmochim Acta 55:3854–3867

Syono Y (1984) Shock-induced phase transition of oxides and silicates. In: Sunagawa I (ed) Materials science of the earth interior I. Terra Science Publications Co., Tokyo, pp 395–414

Syono Y, Goto T, Takei H et al (1981) Association reaction in forsterite under shock compression. Science 214:177

Tomioka N, Fujino K (1997) Natural (Mg, Fe)SiO_3-ilmenite and –perovskite in the Tenham meteorite. Science 277:1084–1086

Tomioka N, Fujino K (1999) Akimotoite, (Mg,Fe)SiO_3, a new silicate mineral of the ilmenite group in the Tenham chondrite. Am Mineral 84:267–271

Tomioka N, Kimura M (2003) The breakdown of diopside to Ca-rich majorite and glass in a shocked H chondrite. Earth Planet Sci Lett 208:271–278

Wang DD (1993) An introduction to chinese meteorites. Science Press, Beijing, pp 101–106 (in Chinese)

Xie XD, Chen M (1997) Shock-produced vapor-grown crystals in Yanzhuang meteorite. Sci China Ser D 40:113–119

Xie XD, Huang WK (1991) Thermal and collision history of the Jilin and Qiangzhen chondrites. Chin J Geochem 10:109–119

Xie XD, Li ZH, Wang DD et al (1991) The new meteorite fall of Yanzhuang—A severely shocked H6 chondrite with black molten materials (Abstract). Meteoritica 26:411

Xie XD, Chen M, Dai CD et al (2001a) A comparative study of naturally and experimentally shocked chondrites. Earth Planet Sci Lett 187:345–356

Xie XD, Chen M, Wang DQ (2001b) Shock-related mineralogical features and P-T history of the Suizhou L6 chondrite. Eur J Miner 13(6):1177–1190

Xie XD, Chen M, Dai CD, El Goresy A, Gillet P (2001c) A comparative study of naturally and experimentally shocked chondrites. Earth Planet Sci Lett 187:345–356

Yagi Y, Suzki T, Akaogi M (1994) High pressure transitions in the system $KAlSi_3O_8$-$NaAlSi_3O_8$. Phys Chem Miner 21:12–17

Zhang J, Herzberg C (1994) Melting experiments on anhydrous peridotite KLB-1 from 5.0 to 22.5 GPa. J Geophys Res 99:17729–17742

Postscript

After we finished writing this book, some new important results in the study of mineralogy of the Suizhou meteorite have been obtained. They are as follows.

1. A shock-induced lithium niobate-structured polymorph of ilmenite $FeTiO_3$ (LN-type phase) has been identified by electron microprobe, Raman spectroscopy, and transmission electron microscopy in the Suizhou chondritic meteorite. This would be the 11th high-pressure phase found in this unique meteorite. The LN-type phase occurs as irregular aggregates of small particles (2–20 μm in size) inside or adjacent to the shock veins. Electron microprobe analyses show that the LN phase and its host ilmenite have nearly identical composition indicating an isochemical solid-state transformation mechanism. Its Raman spectra display bands typical for LN-type phase at 174–179, 273–277, 560–567, and 738–743 cm^{-1}. Crystallographic structure refined from electron diffraction patterns of the LN-type phase indicates a unit cell $a = 5.1241(4)$ Å, $c = 13.733(2)$ Å, and $c/a = 2.68$ in space group R3c.

2. Pyrophanite, $MnTiO_3$, a member of the ilmenite mineral group, has been found in unmelted chondritic rock of the Suizhou meteorite. It was identified by electron microprobe and Raman spectroscopy. It is the third oxide mineral found in this meteorite. Pyrophanite in the Suizhou meteorite occurs in association with troilite and olivine as small grains of irregular shape, i.e., 5–10 μm in size. EPMA results show that this mineral contains 40.87 wt% MnO, 52.04 wt% TiO_2, 4.34 wt% MgO, and 2.83 wt% FeO, yielding a formula $(Mn^{2+}_{0.80}Mg_{0.15}Fe_{0.05})_{1.00}(Mn^{3+}_{0.05}Ti^{4+}_{0.97})_{1.02}O_3$. This indicates that this mineral is close to the end-member pyrophanite.

3. Rutile, TiO_2, was also found in the Suizhou chondritic rock. It was identified by EPMA and Raman spectroscopy. This would be the fourth oxide mineral found in the Suizhou meteorite. Rutile in Suizhou occurs in some ilmenite grains as straight and very thin (1–2 μm in width) lamellae, sometimes only one, crossing the whole grain.

4. Several tens of native copper grains were observed in the Suizhou chondritic rock. This mineral occurs either as small mineral inclusion in kamacite or

© Springer-Verlag Berlin Heidelberg and Guangdong Science
& Technology Press Co., Ltd. 2016
X. Xie and M. Chen, *Suizhou Meteorite: Mineralogy and Shock Metamorphism*,
Springer Geochemistry/Mineralogy, DOI 10.1007/978-3-662-48479-1

taenite, or as individual grains in the interstices of kamacite, taenite, and troilite, and, in rare cases, in the interstices of kamacite, taenite, and silicate minerals. Electron microprobe analyses give the following average contents (in wt%): Cu—92.18, Fe—6.09, Ni—2.68, Co—0.03, and total—100.99.

Printed in the United States
By Bookmasters